Rare Earths and Critical Raw Materials

The Hidden Force Driving Global Power

Top Tier Education Program

ISBN 979-12-243-1433-2

DEDICATION

To those who look beneath the surface of technology and progress.
This work is dedicated to the readers who question, investigate, and connect the hidden structures shaping our world—seeking clarity where others see only complexity.

CONTENTS

ACKNOWLEDGMENTS

Grateful to the researchers, analysts, and industry experts whose work illuminates this complex field.

Thanks to those who build, verify, and challenge data, enabling clearer understanding.

Recognition goes to every reader committed to exploring the strategic realities behind modern technology.

INTRODUCTION
CHAPTER 1 – WHY MATERIALS SHAPE POWER

The story of global power has always been, at its core, a story of materials. Civilizations have risen on the strength of the resources they controlled and fallen when those same resources ran short or lost strategic relevance. Bronze, iron, coal, and oil each defined historical eras not simply because they powered tools and machines, but because they shaped political authority, economic advantage, and military capability. Today, a new category of materials is quietly assuming that same importance: the metals, minerals, and elements underpinning advanced technologies. Rare earths, lithium, cobalt, nickel, gallium, germanium, and dozens of other critical raw materials form the invisible infrastructure of the modern world. These materials, often obscure to the general public, are now central to understanding power in the twenty-first century.

The shift from fossil fuels to electrification and digitalization has intensified the strategic significance of materials in ways previous industrial transformations never required. Unlike oil, which was relatively uniform and could be sourced from many places once extraction technology matured, today's materials landscape is defined by extreme concentration. Single countries sometimes dominate extraction, refining, or components of the value chain to a degree unimaginable in earlier resource eras. China refines the majority of the world's lithium and cobalt, produces over 70 percent of rare earths, and dominates the supply of gallium, germanium, graphite, and several key battery intermediates. The Democratic Republic of Congo supplies most of the world's cobalt. Indonesia has become the epicenter of nickel production. These chokepoints are not abstractions: a disruption in one country can cascade instantly through global manufacturing, halting everything from electric vehicle production to semiconductor fabrication.

Materials also shape power because modern technologies depend on extremely specific physical and chemical properties that cannot be easily substituted. It is not possible to build high-performance electric motors without strong permanent magnets made from neodymium and dysprosium. High-speed telecom networks rely on erbium-doped fibers and lasers fabricated from gallium and indium. Quantum computers need niobium and helium-3. Defense systems—from radar to night-vision optics to missile components—use gallium nitride, germanium, titanium alloys, and rare earth metals in ways that have no quick or easy alternatives. These dependencies create both capability and vulnerability: nations that control the supply can shape technological trajectories, while those that depend on imports face strategic exposure.

The central role of materials becomes visible when examining the energy transition. Clean energy systems require far more metals and minerals than fossil-fuel systems because renewable technologies convert energy using devices—batteries, magnets, semiconductors—rather than burning fuel. Wind turbines require rare earth magnets. Solar panels consume silver, silicon, and specialty dopants. Electric vehicles require lithium-ion batteries with large quantities of lithium, nickel, cobalt, manganese, and graphite. Even the grids needed to transport renewable electricity rely heavily on copper, aluminum, and steel. The transition therefore transforms geopolitics: power no longer depends on pipelines and shipping lanes alone, but on supply chains stretching across continents and dominated by a handful of actors.

Materials influence power not just through scarcity or concentration, but through the structure of supply chains. Most critical materials travel through long, multi-step global routes before becoming part of a finished product. Lithium extracted in Australia may be refined in China, processed into cathode material in Korea, assembled into cells in Europe, and installed in a car in the United States. A disruption at any point—an export ban, a factory closure, a power shortage—can halt the entire chain. These vulnerabilities are amplified by just-in-time manufacturing practices that leave companies with minimal inventories. The lesson has already been demonstrated: when China restricted rare earth exports to Japan in 2010, manufacturers scrambled to find alternative supplies and rapidly adjusted their designs to reduce consumption. The episode revealed that materials can be wielded as strategic instruments, capable of shifting diplomatic behavior and industrial strategy.

The environmental and social dimensions of materials further shape global power dynamics. Many proposed mining and refining projects in democratic countries face public resistance due to concerns over pollution, land use, or community impact. These delays slow diversification efforts, unintentionally reinforcing the dominance of producers with looser regulatory environments. Meanwhile, energy itself intersects with materials: the mining and refining of critical raw materials consume large amounts of electricity, and disruptions in energy supply—such as the electricity curbs that affected Chinese magnesium production in 2021—can instantly translate into shortages of downstream metals. Energy security and material security have become deeply entangled, creating feedback loops that policymakers must now manage together.

Economic power is also fundamentally tied to materials. Entire industries are built on access to specific elements. The competitiveness of electric vehicle manufacturers hinges on secure supplies of lithium, nickel, and graphite. Semiconductor companies depend on specialty gases, high-purity silicon, gallium, and germanium. The digital economy rests on data centers full of capacitors, connectors, chips, and magnets that collectively require dozens of strategic materials. As emerging technologies expand—AI accelerators, quantum computing, advanced robotics—their materials requirements expand as well. Nations that secure steady, affordable access will lead in innovation; those facing shortages will lag, regardless of talent or investment.

This reinforces a broader historical pattern: technological revolutions are always materials revolutions. The Bronze Age depended on tin and copper. The Iron Age depended on iron ore and carbon. The Industrial Revolution relied on coal, steel, and copper. The petroleum age revolved around oil. Our current digital-electrified-automated era is built on a complex periodic table of elements that enable performance, miniaturization, efficiency, and durability. Unlike earlier eras, however, the modern economy depends on dozens of materials simultaneously. Losing access to even one can stall progress across entire sectors.

Another reason materials shape power is the time required to build supply. New mines often take a decade or more to reach production due to permitting, technical challenges, and capital requirements. Processing facilities require expertise and environmental approvals that only a few countries currently possess. Recycling can help but cannot scale quickly enough to meet near-term demand. This lag creates structural scarcity: demand for many materials is rising far faster than supply can expand. Nations that act early to secure long-term partnerships, invest in refining capacity, or build recycling ecosystems will shape global manufacturing patterns decades into the future.

Materials also influence diplomatic relationships. Countries rich in key resources—lithium in South America, cobalt in Africa, nickel in Southeast Asia—have gained new leverage. They can negotiate better terms, require value-added processing domestically, or use export controls to influence global industry. Meanwhile, consuming nations are forming alliances, signing strategic agreements, and building stockpiles to reduce exposure. A new form of geopolitical competition is emerging—not over territory, but over processing plants, mining rights, trade agreements, and technological influence. This competition is subtler than past resource rivalries, but no less consequential.

The strategic importance of materials becomes even clearer when considering defense. Modern militaries rely heavily on advanced materials for sensors, communication systems, engines, armor, and aerospace components. A shortage of gallium could weaken radar systems. Limited germanium supply could affect night-vision capabilities. Constraints in titanium or rare earths could impact aircraft manufacturing. Because substitutes are limited and supply chains are highly concentrated, securing access has become a national security priority for many governments. Stockpiles, strategic reserves, and domestic processing initiatives are re-emerging as tools of statecraft.

In the age of critical raw materials, power is no longer measured only in territory or capital, but in access, control, and resilience across material supply chains. These elements sit at the intersection of technology, industry, and sovereignty, quietly redefining the constraints within which states and corporations operate. Their growing strategic weight is not accidental, but the outcome of deeper structural shifts that have transformed ordinary resources into geopolitical leverage.

CHAPTER 2 – THE AGE OF CRITICAL RAW MATERIALS

The twenty-first century has entered a new age, one defined not by a single dominant resource but by a constellation of minerals, metals, and elements essential for advanced technologies. This is the age of critical raw materials, a period in which economic strength, technological progress, and geopolitical leverage increasingly depend on securing access to substances that, until recently, were unfamiliar outside of specialized industrial and scientific circles. The rise of lithium, cobalt, nickel, rare earth elements, gallium, germanium, graphite, and dozens of other inputs marks a profound shift in the structure of global power. These materials have become indispensable to clean energy, digital infrastructure, aerospace, defense systems, and the manufacturing of high-performance goods. As their strategic value has grown, so has the awareness that their supply chains are fragile, geographically concentrated, and deeply intertwined with global politics.

The emergence of this new age is the result of overlapping technological, environmental, and economic forces. Foremost among them is the global transition toward electrification and renewable energy. Modern clean technologies require far more minerals per unit of energy produced than fossil-fuel systems. Electric vehicles depend on batteries made from lithium, nickel, cobalt, manganese, and graphite. Wind turbines rely on permanent magnets containing neodymium and dysprosium. Solar panels require high-purity silicon, silver, and dopants such as phosphorus or boron. Expanding grids require copper and aluminum at unprecedented scales. Even hydrogen technologies, widely promoted as a complementary pillar of decarbonization, depend on platinum group metals for fuel cells and electrolyzers.

As the world accelerates toward climate goals, the demand for these materials has risen dramatically. Some projections estimate that lithium demand could grow several-fold by 2040, nickel demand could double or triple, and rare earth demand for magnets could grow sharply as electric mobility and wind power expand. This surge has revealed a structural challenge: supply has not kept pace. Mining projects are slow to develop, environmental approvals take years, refining capabilities are concentrated in a handful of countries, and recycling remains at an early stage. In many cases, demand curves are rising on timelines far shorter than the timelines required to bring new supply online. The result is a sustained gap between what the world needs and what it can readily obtain.

The age of critical raw materials is also characterized by extreme geographic concentration. Some minerals are abundant, but economically viable deposits are not. Others are common but require complex, energy-intensive refining that only a few countries have mastered. China plays an outsized role at nearly every stage of many supply chains, from mining and refining to component manufacturing. It processes the majority of global cobalt and lithium, dominates rare earth separation, and controls much of the output of gallium, germanium, graphite, and several battery precursors. The Democratic Republic of Congo produces most of the world's cobalt, Indonesia has become the epicenter of nickel production, and South Africa and Russia dominate the supply of platinum group metals. These patterns mean that global industries—from automotive to electronics to aerospace—depend heavily on a small number of jurisdictions, each influenced by political, environmental, and economic risks.

Supply chain fragility has become one of the defining themes of this new era. Critical raw materials often travel through long, complex global pathways before becoming part of a finished product. Lithium mined in Australia may be refined in China, turned into cathode materials in South Korea, assembled into battery cells in Europe, and installed into electric vehicles in the United States. A single disruption at any point—an export restriction, a strike, a factory shutdown, a power shortage—can ripple across continents. This fragility was demonstrated starkly during the 2020–2022 supply chain disruptions, when energy shortages in China forced the closure of magnesium smelters, triggering downstream shortages of aluminum alloys in Europe. Similarly, the rapid price spikes in lithium, cobalt, and nickel during this period revealed how tight and inflexible these markets truly are.

The age of critical raw materials is also shaped by technological change beyond clean energy. The digital economy depends on a broad set of niche inputs. High-performance semiconductors require gallium, indium, germanium, silicon carbide, and rare specialty gases such as neon and krypton. Fiber-optic networks rely on erbium for optical amplifiers and germanium-doped silica for fiber cores. Quantum technologies depend on materials like niobium, aluminum, helium-3, indium, and rare earth ions. Data centers require copper, aluminum, lithium batteries for backup systems, and high-purity materials for power semiconductors. The proliferation of AI hardware has increased demand for gallium nitride and high-quality silicon carbide, reinforcing the dependence of the digital infrastructure on hard-to-source materials.

Defense and aerospace technologies magnify these dependencies even further. Modern aircraft use titanium alloys for structural components, while satellites rely on gallium arsenide and germanium for high-efficiency solar cells. Night-vision devices require germanium lenses. Advanced radars depend on gallium nitride. Submarine nuclear reactors rely on zirconium and hafnium. Precision-guided munitions need tungsten. Rare earth elements are embedded throughout defense systems, from actuators to guidance units to communication hardware. These materials often come from limited global sources, and the inability to secure stable supplies poses direct risks to national security.

The environmental and social dimensions of critical materials also shape this new age. Mining and refining can be environmentally intensive, generating waste, carbon emissions, and local ecological impacts. Public resistance in democratic countries has slowed diversification efforts. Communities have opposed rare earth processing plants, lithium brine projects, and other facilities due to concerns about pollution, water use, or land disruption. These delays push supply chains toward countries with fewer regulatory constraints, reinforcing the concentration the world is trying to solve. At the same time, cleaner mining and refining technologies are emerging, and they have become strategic priorities for governments seeking both environmental integrity and supply resilience.

Recycling and circularity offer long-term solutions but cannot meet near-term demand. Many critical materials are used in dispersed products—electronics, batteries, motors—that have long lifespans. Large-scale recycling of lithium-ion batteries is only beginning, though it promises to supply an increasing share of cobalt, nickel, and lithium as the volume of end-of-life batteries rises. Rare earth magnet recycling remains technologically complex but is progressing. However, recycling cannot replace primary mining for decades, especially for rapidly growing markets where installed stock is still small compared to future demand.

The financial dimension of critical raw materials has also intensified. Price volatility has become a hallmark of these markets, driven by tight supply, speculative investment, policy signals, and technological shifts. Lithium prices surged dramatically in 2021–2022 before falling again as new supply entered the market and demand fluctuated. Nickel experienced extreme volatility, at one point triggering unprecedented market interventions. These swings complicate investment planning and create uncertainty for manufacturers dependent on stable input costs. At the same time, financial interest in "energy transition metals" has grown, attracting hedge funds, sovereign wealth funds, and specialized mining investment vehicles. This capital can accelerate project development but can also contribute to speculative cycles.

Government policy has emerged as one of the most influential forces in this new age. Nations have developed critical minerals strategies, stockpiled supplies, formed alliances, and imposed trade restrictions. Export bans on raw materials—such as Indonesia's decision to halt nickel ore exports—have reshaped global industrial patterns, pushing companies to build downstream processing domestically. The United States, European Union, Japan, and South Korea have invested in domestic refining, recycling, and advanced materials manufacturing. New agreements with resource-rich countries aim to secure stable, long-term access while promoting local development. Multilateral initiatives have formed to monitor supply risks, coordinate policy responses, and support sustainable mining practices.

The age of critical raw materials is defined not only by scarcity and concentration but by interdependence. Technologies that represent the future of economic growth depend on materials that come from complex, vulnerable global systems. Substitution efforts can help reduce pressure—for example, the rise of lithium iron phosphate batteries reduces cobalt and nickel demand, while research into magnet-free motors may reduce rare earth dependence—but these innovations often shift rather than eliminate criticality. Replacing one material frequently creates new dependencies on others. Entire industries are optimized around the performance characteristics of specific materials, and redesigning these systems takes time, investment, and technical breakthroughs.

As the world moves deeper into this era, nations must navigate a delicate balance: expanding supply while minimizing environmental harm, securing access while avoiding destabilizing competition, and innovating new technologies while managing the risks of substitution. The age of critical raw materials is not simply about geology or engineering; it is about reshaping economic models, industrial strategies, and geopolitical priorities.

What defines the current phase is not a single driver, but the compression of multiple systemic pressures into the same strategic space. Energy systems, digital infrastructures, military capabilities, and industrial resilience now draw from the same finite material base. As these pressures accumulate, resources once treated as technical inputs become instruments of influence and constraint. The logic of international relations is increasingly shaped by this material reality.

CHAPTER 3 – TECHNOLOGY, GEOPOLITICS, AND THE NEW RESOURCE RACE

The rise of critical raw materials has not only reshaped industries and technological trajectories; it has fundamentally altered the logic of geopolitics. As the global economy becomes more electrified, digitalized, and automated, the materials enabling these systems have become instruments of power. Nations now compete not just for markets or influence, but for control over the elements that make advanced technologies possible. The result is a new resource race, one far more complex and multidimensional than the oil rivalries of the twentieth century. It is a race fought through supply chains, investment strategies, export controls, alliances, technological innovation, and economic statecraft. Understanding this emerging landscape is essential for grasping the strategic realities of the twenty-first century.

The interplay between technology and geopolitics is the starting point. Modern technologies increasingly rely on specialized materials with unique properties—magnetic, luminescent, thermal, structural, catalytic—that cannot be easily replaced. These properties define the performance of electric motors, semiconductors, batteries, fiber-optic communication systems, radar arrays, satellites, and countless other devices. Because these technologies underpin critical sectors—from energy to defense to digital infrastructure—control over the materials that enable them becomes a source of national leverage.

This dynamic is particularly visible in the context of China's role in global supply chains. Over several decades, China invested heavily in mining, refining, and processing capabilities across a wide range of critical raw materials. It built the world's dominant rare earth separation industry, captured processing stages for lithium, cobalt, and graphite, and secured access to overseas mines through state-backed investments. It also established leadership in battery precursor materials, permanent magnets, and advanced components necessary for clean energy and electronics. This position gives China structural leverage: disruption or restriction in any of these supply chains can immediately impact manufacturers worldwide.

Export controls have already demonstrated their power. When China restricted rare earth shipments to Japan in 2010 over a diplomatic dispute, the incident revealed how quickly supply chains could be weaponized. In recent years, export licensing requirements for gallium and germanium sent shockwaves through semiconductor and optics industries, highlighting how dependent advanced economies are on specific materials that come from only a few sources. These actions crystallized a geopolitical reality: materials once considered obscure and technical have become central to national security, economic resilience, and the balance of power.

Other nations have responded by reassessing their own vulnerabilities. The United States, European Union, Japan, South Korea, and others have launched strategies to reduce dependence on concentrated suppliers. These measures include building domestic refining and processing facilities, investing in recycling technologies, supporting new mines, and forming partnerships with resource-rich countries. For example, efforts to develop supply agreements with nations in Africa, South America, and Southeast Asia aim to secure diversified access while supporting local value creation. Yet diversification is not straightforward. It requires long-term investment, technical expertise, regulatory capacity, and stable political environments. It also requires navigating local environmental and social concerns that often slow mining projects in democratic countries.

The new resource race is not simply about extraction; it is equally about the ability to refine, process, and convert materials into high-value components. Processing capacity is often far more concentrated than mining. Even when countries possess mineral deposits, they may lack the industrial ecosystem needed to move beyond raw ore. Processing rare earth elements into usable oxides and metals, converting lithium into battery-grade chemicals, or producing high-purity silicon and compound semiconductors requires specialized facilities and expertise that only a few countries have. Building this capability takes years and enormous capital. Nations that control refining and processing therefore hold disproportionate power, regardless of where the raw minerals are mined.

Resource nationalism has also re-emerged as a geopolitical force. Countries with rich mineral endowments increasingly seek to capture more value domestically, tighten control over exports, or reshape global supply chains to their advantage. Indonesia's ban on raw nickel ore exports, intended to force companies to build local processing plants, dramatically reshaped global nickel supply. Chile and Bolivia have pursued new approaches to lithium governance, emphasizing state participation and stricter conditions for foreign investors. African nations are increasingly scrutinizing past agreements and pushing for fairer terms, local processing requirements, or partnerships that share benefits more equitably. These shifts reflect a broader trend: resource-rich nations are no longer willing to remain at the bottom of global value chains.

Competition extends to financial and diplomatic arenas as well. Major powers are investing diplomatically in regions with strategic resources, offering infrastructure, financing, and security assistance to secure influence. Mining concessions in Africa, lithium projects in South America, rare earth deposits in Australia and Southeast Asia, and copper resources in the Andes have become the focus of intensified geopolitical attention. Western nations now seek to counterbalance China's longstanding investments in these regions, often appealing to transparency, environmental standards, and sustainable development. Meanwhile, emerging powers view critical minerals as levers to attract industrial investment or forge new alliances.

The geopolitical significance of critical materials is amplified by their role in national security. Defense systems—from advanced radar to stealth aircraft to guided weapons to secure communications—depend on specific elements that are sourced from limited suppliers. A shortage of gallium could impede radar production; limited germanium supply could affect night-vision systems; restricted rare earth availability could delay the manufacturing of precision-guided munitions or electric naval systems. Nations therefore treat these materials not merely as economic inputs but as strategic reserves. Stockpiling, long-term contracts, and government-backed investment in domestic supply chains become tools of defense planning.

Technology competition further intensifies this resource race. As countries pursue leadership in artificial intelligence, quantum computing, advanced manufacturing, and clean energy, they compete for the materials that enable these capabilities. AI data centers need advanced semiconductors built with gallium and silicon carbide; quantum computers need niobium, aluminum, and helium isotopes; high-efficiency solar panels rely on silver, tellurium, and indium; electric vehicles depend on lithium-ion batteries. The more advanced the technology, the narrower the set of materials it requires—and the greater the geopolitical contest to secure them.

Yet competition is only part of the story. Cooperation also plays a role, particularly among nations seeking to stabilize markets, reduce volatility, and promote responsible sourcing. Multilateral initiatives aim to coordinate supply assessments, share information, support mining investments, and develop best practices for environmental and social standards. These initiatives emphasize that the world does not face a shortage of minerals in the Earth's crust, but rather a shortage of investment, time, and politically acceptable projects. International collaboration therefore becomes a mechanism to accelerate the development of diversified, transparent, and sustainable supply chains.

Still, cooperation has limits. As long as the global economy depends on concentrated supplies and high-growth technologies, competition will remain a defining feature. Export controls, industrial subsidies, strategic stockpiles, and alliance-building efforts will continue to shape the landscape. The risk is that the new resource race could become a source of geopolitical tension, particularly if nations perceive supply disruptions as deliberate acts of coercion. Alternatively, poorly managed competition could slow the energy transition, worsen environmental impacts, or exacerbate inequality between resource-rich and resource-poor nations.

A key challenge in this emerging geopolitical environment is the tension between sustainability and strategic urgency. Nations want to scale up supply quickly to meet climate goals and technological needs, but they also face pressure to ensure that mining and refining are clean, ethical, and socially responsible. This tension can slow diversification efforts, reinforcing dependence on countries that operate under weaker environmental and labor standards. At the same time, the push for circularity—through recycling, urban mining, and materials recovery—offers long-term potential to ease supply pressure, though it will take decades to reach maturity.

The new resource race is therefore a race on multiple fronts: extraction, processing, technological substitution, recycling, diplomacy, finance, innovation, and sustainability. It is a race defined by interdependence, asymmetry, and rapid change. It is also a race with no single winner. In an interconnected world, no nation can secure all stages of every supply chain. Instead, resilience will emerge from diverse partnerships, investment in innovation, and governance that balances national interests with global stability.

PART I – FOUNDATIONS: UNDERSTANDING RARE EARTHS AND CRITICAL MATERIALS

CHAPTER 4 – WHAT RARE EARTHS REALLY ARE

Rare earths occupy a paradoxical place in the modern world. They are at once indispensable and obscure, technologically vital yet geologically misunderstood, strategically important but often invisible to the people whose lives depend on them. They underpin the most advanced tools of the twenty-first century, from electric vehicle motors to fiber-optic networks, yet they remain hidden behind a name that confuses almost as much as it informs. To understand the broader landscape of critical raw materials, we must begin by clarifying what rare earths actually are, how they behave, why they matter, and how their unique properties have transformed the world.

The term "rare earths" refers to a group of 17 metallic elements: the 15 lanthanides on the periodic table, plus yttrium and scandium. Chemically, they are remarkably similar to each other. Most are trivalent metals with comparable ionic radii, which means they often occur together in the same mineral deposits. This similarity makes them difficult to separate, one of the reasons their supply chain is technically complex and highly specialized. Despite their name, rare earths are not particularly rare in terms of crustal abundance. Cerium, for example, is more abundant than copper. The real scarcity lies not in their presence on Earth but in the availability of economically concentrated deposits and in the ability to refine them into pure, usable forms. Their extraction and separation involve multi-step chemical processes that are costly, technically demanding, and environmentally sensitive.

Rare earths are commonly grouped into two categories: light rare earth elements (LREEs) and heavy rare earth elements (HREEs). The distinction is based on atomic weight but has major implications for supply and applications. Light rare earths include elements like lanthanum, cerium, praseodymium, and neodymium. These tend to be more abundant geologically and easier to extract. Heavy rare earths—including europium, terbium, dysprosium, and others—are rarer, often found in smaller quantities, and significantly more challenging to separate. This difference in availability directly affects their market value and strategic importance. Heavy rare earths, for instance, play a crucial role in high-performance permanent magnets and advanced defense technologies, making them some of the most sought-after elements in modern supply chains.

Each rare earth element possesses a unique combination of magnetic, luminescent, optical, and electrochemical properties. These are not interchangeable traits; they stem from the elements' unfilled 4f electron orbitals, which give rise to behaviors unmatched by other metals. Neodymium, praseodymium, and dysprosium, for example, are central to the creation of neodymium-iron-boron magnets—the strongest permanent magnets known. These magnets are integral to the motors in electric vehicles, the generators in offshore wind turbines, and the miniaturized components in consumer electronics.

Their exceptional strength allows engineers to create lighter, more efficient, and more compact systems, enabling breakthroughs in electrification and miniaturization.

Other rare earths shine—literally—in different ways. Europium and terbium are responsible for the red and green colors in fluorescent lighting and LED displays. Yttrium combines with aluminum and other rare earths to form garnet materials used in lasers and microwave filters. Cerium serves as a glass-polishing compound and a catalyst, appearing in everything from camera lenses to automotive catalytic converters. Lanthanum improves the quality of optical glass, enabling the lenses used in smartphones and professional cameras. Gadolinium's neutron-absorbing properties make it vital for nuclear reactor control rods and MRI contrast agents. In short, rare earths are enabling materials: their properties make possible technologies that would otherwise be impractical or entirely nonfunctional.

Where rare earths differ most from many other minerals is in the structure of their supply chain. While deposits exist in many countries, economically viable mining operations are comparatively few. More importantly, the refining and separation stages are highly concentrated. The difficulty lies in the chemical similarity of the elements: separating them requires elaborate solvent-extraction processes involving hundreds of stages, large volumes of reagents, and careful environmental management. This complexity contributed to the global consolidation of refining capacity in a few countries. Once separation is complete, rare earth oxides must be processed into metals and then into magnet alloys, phosphors, or other advanced materials—steps that further narrow the list of countries involved.

Another complexity arises from the fact that rare earths often appear as byproducts of mining for other metals. Some deposits contain high concentrations of light rare earths but few heavy ones. Others produce heavy rare earths but in smaller volumes, making them scarce and expensive. In many cases, the balance of supply and demand depends not on geology but on processing capacity, market structure, and the economics of byproduct extraction. For example, cerium and lanthanum are abundant and sometimes oversupplied, whereas dysprosium and terbium, essential for high-temperature magnets in electric motors and wind turbines, remain in chronic short supply.

The importance of rare earths has grown in tandem with the rise of clean energy and digital technologies. Electric vehicles rely on neodymium-based magnets for their traction motors, especially in high-performance models. Wind turbines use magnets made with neodymium and dysprosium in their direct-drive generators, enabling more compact designs that are ideal for offshore installations. Smartphones use rare earths in speakers, cameras, vibration motors, and phase-shifting components. Fiber-optic networks depend on erbium, which amplifies light signals across transoceanic distances. Even emerging fields such as quantum computing rely on rare earths in the form of doped materials used for advanced sensing, lasers, or storage.

The strategic importance of rare earths is heightened by the fact that demand is rising rapidly while supply chains remain tightly constrained. As electric vehicle adoption accelerates, motor production increases dramatically. As renewable energy systems expand, magnet demand rises accordingly. Yet scaling rare earth supply is far slower than scaling manufacturing. New mining projects must overcome environmental, regulatory, and financial hurdles. New refining capacity requires specialized technology and experienced personnel. Environmental concerns may delay or prevent projects in many countries, pushing supply chains toward jurisdictions with more permissive regulations. The result is a structural imbalance between supply and demand that is likely to persist for years.

Environmental considerations are central to the rare earth story. Mining and processing generate waste, including radioactive byproducts in some cases, and require substantial water and chemical usage. These impacts have drawn criticism when not managed properly. However, modern technologies and improved environmental standards have significantly reduced the footprint of responsible rare earth production. Cleaner refining methods, advanced waste treatment, and better monitoring are gradually transforming the industry. Still, the environmental consequences of past extraction have shaped public perception, contributing to resistance against new projects in some regions. This tension between environmental responsibility and strategic necessity is one of the defining challenges of rare earth supply.

Rare earths also reveal the fragility of globalized supply chains. Their distribution from mine to magnet spans continents and involves numerous industries. A single bottleneck—whether a processing facility shutdown, a geopolitical dispute, or an energy shortage—can disrupt entire sectors. For example, power rationing in key producing regions has previously caused shortages of upstream materials, which in turn reduced the availability of downstream products like magnets or phosphors. As technologies become more dependent on rare earths, these vulnerabilities become more acute.

The strategic dimension of rare earths extends beyond supply risk. Their role in defense technologies amplifies their importance to national security. Guided missiles use rare earths in sensors and actuators, advanced aircraft rely on them in control systems and engines, and secure communication devices depend on rare earth components. Satellites incorporate rare earth-based lasers and optical materials. A shortage of critical rare earth elements would not only impact commercial industries but could compromise defense readiness and technological superiority.

Yet the story of rare earths is not entirely one of vulnerability. It is also one of innovation and adaptation. Research into new magnet chemistries aims to reduce or eliminate reliance on heavy rare earths. Advances in motor design, such as switched-reluctance and wound-field motors, offer alternatives that use fewer or no permanent magnets. In battery technologies, chemistries like lithium iron phosphate reduce demand for nickel and cobalt, though rare earths remain essential for motors and other components. Recycling technologies for rare earth magnets, though still maturing, are expanding. These developments suggest that while rare earths will remain indispensable, the world is exploring ways to manage and optimize their use.

Understanding rare earths also illuminates broader patterns in critical raw materials. They exemplify the challenges of complex supply chains, the role of advanced materials in technological progress, the tension between environmental stewardship and industrial necessity, and the geopolitical dynamics that arise when essential inputs are concentrated in a few regions. They serve as a microcosm of the entire field of critical materials: abundant yet scarce, essential yet underappreciated, technically challenging yet central to modern life.

CHAPTER 5 – CRITICAL RAW MATERIALS: DEFINITIONS AND CRITERIA

The concept of critical raw materials has become central to understanding how economies, technologies, and geopolitical systems function in the twenty-first century. Yet for all its prominence, the term "critical" is often used without a clear sense of what it actually means. It does not simply refer to scarcity, nor does it refer only to strategic importance. A material becomes "critical" through a combination of economic necessity, supply vulnerability, technological dependence, and the difficulty of replacing it. In this chapter, we define what critical raw materials are, examine the criteria used to classify them, and explore why they have become indispensable to the global economy.

Critical raw materials—often abbreviated as CRMs—refer to mineral or elemental resources that are essential to major industries or national security, but whose supply is threatened by concentration, geopolitical risk, or other structural vulnerabilities. The concept is inherently dynamic: what is considered critical today might not be critical tomorrow, either because new technologies reduce demand or because new sources of supply emerge. Conversely, materials once considered unremarkable may become critical as new applications arise. The rise of electric mobility, renewable energy, and semiconductor-rich digital systems has rapidly elevated elements such as lithium, cobalt, nickel, and gallium from niche commodities to globally strategic inputs.

The most widely accepted definitions of critical raw materials share two core dimensions: economic importance and supply risk. Economic importance reflects how essential a material is to key sectors of the economy, especially those that are high-growth, high-value, or essential for societal functioning. Supply risk captures vulnerability to disruption, which can stem from geological concentration, political instability, environmental barriers, market structure, or processing bottlenecks. A material becomes critical when high economic importance intersects with high supply risk.

Understanding this intersection requires looking at how modern technologies depend on specific materials. In clean energy, lithium is essential to high-performance batteries, cobalt provides stability and energy density, nickel enhances capacity, graphite forms the anode structure, and rare earth elements enable high-efficiency motors and wind turbines. In electronics, gallium, germanium, indium, tantalum, and rare earths underpin semiconductors, optical systems, smartphones, and data centers. In aerospace and defense, titanium, niobium, tungsten, rare earths, and platinum group metals are indispensable. Because these sectors are foundational to technological advancement, their material inputs acquire strategic importance.

Supply risk arises from multiple sources, and rare earths illustrate several of them simultaneously. Geological concentration is one major factor: while rare earths are not scarce in absolute terms, economically extractable deposits are limited. Processing capacity is even more concentrated, requiring complex chemical separation technologies that only a few countries have developed at scale. Many other materials exhibit similar patterns. Over 70 percent of cobalt ore production comes from a single region in the Democratic Republic of Congo. Indonesia produces more than half of global nickel. South Africa and Russia dominate platinum group metals. China is the leading producer or processor of numerous materials, including rare earths, graphite, gallium, germanium, and much of the world's lithium and cobalt refining.

A second source of supply risk is the structure of global supply chains. Critical materials often travel through several countries from mine to finished component. Lithium extracted from brine in South America may be processed into battery chemicals in China, turned into cathode materials in Japan or Korea, and assembled into battery cells in Europe or the United States. Each step represents a potential point of failure. A disruption in one country—caused by export controls, political instability, drought, strikes, or energy shortages—can ripple through the entire chain. The magnesium shortage linked to energy rationing demonstrated how unexpectedly vulnerable supply chains can be, even for materials not widely recognized as strategic.

A third element of supply risk is the long lead time for new production. Opening a new mine can take a decade or more due to exploration, permitting, financing, environmental reviews, and community negotiations. Refining plants require specialized expertise and costly infrastructure. Even when governments identify materials as critical, expanding supply is slow and complicated. Meanwhile, demand for these materials is accelerating quickly as electric vehicles scale, renewable energy systems expand, and digital infrastructure grows. This mismatch between rising demand and slow supply response increases the likelihood of shortages, price volatility, and geopolitical tension.

Environmental and social factors add further complexity. Many critical raw materials are extracted in regions facing governance challenges, environmental degradation, or social conflict. Cobalt mining in the Congo, for instance, has raised concerns about labor practices and environmental impacts. Lithium extraction from brines in arid regions raises questions about water usage and local ecosystems. Rare earth processing can generate hazardous waste if not managed properly. These issues affect public perception, regulatory decisions, and investment flows, and they influence how readily countries can diversify their supply chains.

Another dimension of criticality is substitutability, or the ease with which a material can be replaced by others. In many cases, substitutes exist in theory but not in practice without sacrificing performance, efficiency, size, or cost. For example, neodymium magnets could be replaced by ferrite magnets, but ferrites are significantly weaker, heavier, and less efficient. Cobalt in batteries can be reduced but not eliminated entirely in many chemistries, except for certain types like LFP, which have lower energy density. Gallium in high-frequency semiconductors has unique properties that silicon cannot replicate. Even when substitutes are available, switching requires redesigning products, retooling factories, and developing supply chains for alternative materials. As a result, substitutability tends to be low for most advanced technologies.

The criteria used to identify critical materials vary by institution but generally revolve around these same themes. The United States, European Union, and International Energy Agency each publish their own lists, and although they differ in emphasis, they share a consistent analytical framework. Economic importance, supply concentration, trade dependencies, environmental vulnerability, and strategic relevance shape which materials are included. Some institutions also consider import reliance and the potential for co-production or byproduct extraction. The classification is ultimately a policy tool, aimed at guiding investment, regulation, and strategic planning.

Criticality is not static. As industries evolve, supply chains reorganize, and technologies mature, the status of materials can change. Nickel became more critical with the rise of high-nickel battery chemistries. Cobalt's criticality peaked as EV production expanded, but may diminish if new chemistries reduce reliance on it. Rare earth demand has surged as electrification grows, especially for dysprosium and terbium in high-performance magnets. Copper, once considered abundant, has entered discussions of criticality due to and its centrality in electrification and constraints on new supply. Even aluminum, historically well-supplied, appears on some critical lists because of energy-intensive refining and production concentration.

One of the most interesting aspects of critical materials is their broadness. They encompass not just exotic elements with complex names, but also everyday industrial metals whose significance is rising due to new applications. Copper, nickel, manganese, and graphite illustrate this point. Their volumes are much larger than those of rare earths or gallium, but they are no less essential to the energy transition. At the same time, small-volume materials like germanium or hafnium can shape entire industries despite being produced in tiny quantities.

Criticality also highlights a deeper structural reality: advanced economies depend on a periodic table of materials that are unevenly distributed across the globe. Unlike the fossil-fuel era, when transportable commodities like oil could be accessed through global markets, the new materials landscape involves dozens of inputs, many with highly specialized supply chains. This dispersal makes resilience more challenging. No single country controls the entire chain, yet many countries depend on specific bottlenecks.

In this landscape, policy becomes a critical factor. Governments deploy a range of tools to mitigate supply risks: strategic stockpiles, subsidies for domestic processing, support for recycling industries, public-private partnerships for new mining projects, and diplomatic engagement with resource-rich countries. Export controls, investment screening, and environmental standards all play roles in shaping access and influence. As more countries adopt critical minerals strategies, the competition for investment, processing capacity, and long-term contracts intensifies.

Recycling and circularity represent one of the few long-term opportunities to reduce dependence on primary extraction. Many critical materials can, in principle, be recovered from end-of-life products, though technical and economic barriers remain significant. Battery recycling can yield lithium, cobalt, nickel, and manganese, reducing pressure on mining. Magnet recycling can reclaim neodymium and dysprosium. Semiconductor recycling can extract gallium and indium. However, because most of the products containing these materials are still early in their lifespan, large-scale recycling will not significantly ease supply constraints for at least another decade.

Substitution research is another pillar of criticality mitigation. Scientists and engineers are exploring magnet alternatives that reduce heavy rare earth usage, new battery chemistries that rely on abundant materials, and semiconductor technologies less dependent on gallium or germanium. These innovations offer promise, but adoption is slow and often limited by performance trade-offs. Substitution seldom eliminates criticality fully; it usually shifts it from one material to another.

The concept of critical raw materials therefore serves as both a diagnostic tool and a strategic framework. It clarifies where vulnerabilities lie, what technologies depend on, and how nations must adapt. It also illustrates the interconnected nature of modern systems: economic resilience, technological innovation, environmental sustainability, and geopolitical stability are all influenced by the steady flow of materials that most people never see.

Understanding what makes a material critical is essential before exploring their properties, applications, and global distribution. The extraordinary capabilities of modern technologies are rooted in the specific chemical and physical properties of these materials. Those same properties impose rigid constraints on substitution, limiting flexibility across industrial and strategic systems. What appears as a technical detail at the atomic level translates into leverage and vulnerability at the global scale.

CHAPTER 6 – PROPERTIES AND APPLICATIONS ACROSS MODERN TECHNOLOGY

Modern technology is built on a foundation of materials whose properties are as unusual as they are indispensable. These properties—magnetic, optical, electrical, thermal, catalytic, and structural—determine what devices can do, how efficiently they operate, how small they can become, and how transformative they are for entire industries. Understanding why certain materials are irreplaceable requires examining these properties in detail and exploring how they are deployed across the technologies that define the twenty-first century: electronics, clean energy, digital infrastructure, aerospace, and defense. Critical raw materials are not simply commodities; they are the functional building blocks of the modern world.

To begin, the most strategically significant properties are magnetic. Elements such as neodymium, praseodymium, and dysprosium are central to the strongest permanent magnets ever created: neodymium-iron-boron magnets. These magnets deliver unmatched power density, enabling compact, high-efficiency motors used in electric vehicles, industrial automation, robotics, drones, and wind turbines. Their strength arises from quantum-level electron configurations unique to rare earths, allowing them to maintain magnetic performance even at high temperatures when doped with heavy rare earths like dysprosium or terbium. Without these magnets, electric motors would need to be significantly larger, heavier, and less efficient, undermining the economics of electrification. These magnetic materials also enable the tiny speakers and haptic motors in smartphones and the voice coil actuators in hard disk drives. The miniaturization of countless consumer technologies has been made possible precisely because these rare earth magnets combine strength, temperature stability, and durability in ways no other material can.

Other critical materials serve key optical functions. Erbium, for example, is the element that enables modern global telecommunications. When doped into fiber-optic cables, erbium amplifies signals across thousands of kilometers under the ocean, making high-speed internet possible. Germanium functions as a dopant in fiber cores, adjusting refractive index and enabling light transmission with minimal loss. In infrared optics, germanium and indium antimonide appear in night-vision systems, thermal cameras, and military sensors. These materials absorb and transmit specific wavelengths with remarkable precision. Indium phosphide and gallium arsenide, both compound semiconductors, are used to fabricate the lasers that power high-speed fiber communications. Their direct bandgap properties allow them to emit light efficiently at key wavelengths, enabling data to travel across continents through hair-thin cables. Modern connectivity is therefore not simply a story of software and electronics—it is grounded in the unique optical behaviors of a handful of critical elements.

Electrical properties form another cornerstone of cutting-edge technologies. The semiconductor revolution depends on materials capable of controlling electrical current at nano-scale precision. Silicon remains the foundation of mainstream computing, but many advanced applications require compound semiconductors such as gallium nitride and silicon carbide. Gallium nitride allows transistors to operate at high voltage and high frequency with minimal energy loss, making it essential for 5G telecommunications, radar systems, power electronics in electric vehicles, and high-performance chargers. Silicon carbide is indispensable in power electronics for electric vehicles, renewable energy inverters, and industrial machinery, thanks to its ability to handle high temperatures and voltages with exceptional efficiency. These materials reduce energy losses, improve performance, and enable compact designs that silicon alone cannot achieve.

Beyond semiconductors, myriad critical materials enable the operation of modern electronics. Tantalum's ability to form an extremely thin but stable oxide layer makes it ideal for capacitors in smartphones, laptops, and data centers, where high reliability and miniaturization are essential. Indium-tin oxide coats touchscreens and flat-panel displays, offering transparency and electrical conductivity unmatched by alternative materials. Rare earth phosphors—particularly europium and terbium—generate the vivid colors in LED displays and energy-efficient lighting. Even the wiring and connectors within devices rely on silver, gold, and copper for conductivity and corrosion resistance. The digital world, though often perceived as immaterial, is grounded in a dense network of critical materials that translate electrons into information.

Thermal properties are equally critical. Some technologies operate at extreme temperatures, requiring materials that conduct, resist, or retain heat with precision. Graphite is used not only as a battery anode but also as a high-temperature material in furnaces and reactors due to its thermal stability. Silicon carbide again plays a role, enabling high-temperature semiconductors for harsh-environment electronics. Helium, particularly helium-4 and helium-3, is indispensable for cryogenic cooling in quantum computing, MRI machines, and scientific research. Without these gases, superconducting qubits, ultra-low-temperature sensors, and certain types of optical experiments would be impossible. Meanwhile, alloys based on nickel, chromium, and cobalt provide strength at high temperatures, forming the backbone of jet engines, gas turbines, and industrial equipment. Each of these materials occupies a niche dictated by the physics of temperature—niches that few other elements can fill.

Catalytic properties form another pillar of modern technology. Platinum group metals—platinum, palladium, rhodium, and iridium—are among the most effective catalysts known. They are used in automotive catalytic converters, where they convert toxic exhaust gases into less harmful emissions. They also play central roles in hydrogen technologies: platinum enables fuel cells, while iridium is indispensable for electrolyzers that produce green hydrogen. Cerium and lanthanum appear in catalytic converters and refining processes, enhancing chemical reactions and improving energy efficiency. The importance of catalytic materials will increase as industries shift toward low-carbon production methods, making them integral to the global energy transition.

Structural properties of critical materials underpin aerospace, defense, and advanced engineering. Titanium alloys combine strength, corrosion resistance, and low density, making them essential in aircraft frames, spacecraft, medical implants, and naval systems. Their ability to withstand extreme temperatures and stresses allows them to serve in environments where steel or aluminum would fail. Tungsten, one of the densest metals known, is used in armor-piercing projectiles, aerospace counterweights, and specialized industrial tools. Its hardness and high melting point make it critical for applications requiring exceptional durability. Niobium enhances the performance of superalloys used in jet engines, rockets, and gas turbines, where even slight improvements in heat resistance can dramatically increase efficiency. Hafnium and zirconium play central roles in nuclear technologies, with hafnium absorbing neutrons in control rods and zirconium enabling fuel cladding that resists corrosion. These materials are the mechanical backbone of systems where failure is not an option.

Applications in renewable energy form one of the fastest-growing drivers of critical materials demand. Solar photovoltaics rely on high-purity silicon, silver, and dopant materials. Silicon must achieve extraordinary levels of purity—up to 99.9999 percent—to function effectively as a semiconductor. Silver forms the conductive pathways on solar cells, enabling the extraction of electrical current. Advanced solar technologies use additional critical materials: tellurium and cadmium in cadmium-telluride thin-film panels, or indium, gallium, and selenium in CIGS technologies. Although silicon dominates the market, these thin-film technologies remain important for specific applications and drive demand for their respective materials.

Wind turbines, particularly offshore models, rely heavily on rare earth magnets in their direct-drive generators. These magnets reduce the need for complex gearboxes and improve reliability in harsh environments. As nations expand offshore wind capacity, demand for neodymium, praseodymium, and dysprosium increases accordingly. Meanwhile, the supporting infrastructure for renewable energy—such as grid expansion—requires large quantities of copper and aluminum for transmission lines and steel for towers. Power electronics used in inverters and converters depend on silicon carbide and gallium nitride, illustrating how renewable energy relies on a web of critical materials far more diverse than fossil fuel systems ever required.

Energy storage technologies deepen this dependence. Lithium-ion batteries require lithium, cobalt, nickel, manganese, graphite, copper, and aluminum. Each chemistry emphasizes different materials, but all rely on lithium as the core component. Cobalt ensures stability and extends battery life, nickel increases energy density, manganese improves safety, and graphite—natural or synthetic—forms the anode of nearly every commercial lithium-ion battery. As electric vehicle production grows, demand for these materials scales exponentially. Grid-scale energy storage introduces further materials such as vanadium used in redox flow batteries. Platinum enables hydrogen fuel cells, creating another link between critical materials and energy systems.

Digital infrastructure expands this dependence even further. Data centers require copper, aluminum, silicon, rare earth magnets, tantalum capacitors, high-purity gases, and advanced semiconductors. The rise of artificial intelligence, which depends on high-performance GPUs and specialized accelerators, drives demand for gallium, silicon carbide, gold, and advanced packaging materials. Gallium nitride and silicon carbide are becoming essential for power management in data centers, while gold and palladium appear in connectors and bonding wires. The explosive growth of cloud computing means that even small quantities of critical materials, multiplied across millions of servers, create significant cumulative demand.

Aerospace and defense represent the highest-performance uses of critical materials. Satellites use gallium arsenide solar cells, germanium substrates, beryllium mirrors, and rare earth-based actuators. Missiles and guidance systems incorporate rare earth sensors, tungsten penetrators, and high-performance alloys. Radar systems rely on gallium nitride for high-power radio-frequency transmission. Submarines depend on zirconium-based fuel cladding for nuclear reactors and rare earth actuators for acoustic systems. These applications require materials with properties at the outer limits of physics—materials that must withstand radiation, extreme temperature, pressure, or vibration. Failure is unacceptable, making reliability and purity paramount.

What emerges from this landscape is a clear pattern: modern technology is defined by materials whose properties cannot be easily substituted. The unique electron structures of rare earths, the bandgap characteristics of compound semiconductors, the catalytic behavior of platinum group metals, and the structural capabilities of aerospace alloys collectively enable the technologies that power the world. Substitutions exist in theory, but in practice they often result in inferior performance, increased weight, higher energy consumption, or greater cost. This technological dependence elevates the strategic importance of the materials described throughout this chapter.

CHAPTER 7 – MAPPING THE PERIODIC TABLE OF POWER

Understanding critical raw materials requires more than listing their names or summarizing their applications. It requires reimagining the periodic table itself—not as a neutral scientific chart, but as a geopolitical map, an industrial blueprint, and a reflection of twenty-first-century power dynamics. The modern world does not treat all elements equally. Some elements anchor global supply chains, others shape military capability, others determine the pace of the energy transition, and still others serve as invisible inputs to digital infrastructure. In this sense, the periodic table has become a map of influence, vulnerability, and strategic leverage.

Mapping the periodic table of power begins with recognizing that different families of elements play different roles in the global system. The rare earth elements occupy one of the most prominent positions. As explained earlier, their magnetic, luminescent, and electrochemical properties make them indispensable in electric motors, wind turbines, optical systems, catalysts, and defense technologies. But rare earths are not alone. Adjacent to them on this conceptual map are transition metals such as lithium, nickel, cobalt, copper, nickel, manganese, chromium, vanadium, and molybdenum—each essential to the batteries, alloys, grids, or industrial systems that underpin modern technology. These metals form the backbone of electrification, energy storage, and advanced manufacturing, making them central to the new industrial age.

Another cluster of strategic elements lies within the semiconductors and optics domain—gallium, germanium, indium, tellurium, selenium, and arsenic. These materials enable the compound semiconductors used in high-frequency electronics, LEDs, fiber-optic communication, infrared sensors, and satellite technologies. They are small-volume markets, yet their strategic importance is immense. Without gallium nitride and silicon carbide, for example, high-efficiency power electronics for electric vehicles and data centers would not exist. Without indium phosphide and gallium arsenide, global fiber networks would lose the performance needed to sustain high-speed communication. In this cluster, a shortage of even a few tonnes of material can disrupt entire industries.

The noble metals—platinum group metals (platinum, palladium, rhodium, iridium), as well as silver and gold—occupy another region of the map. Their exceptional catalytic, conductive, and corrosion-resistant properties anchor key technologies. Platinum and iridium enable hydrogen production and fuel cells. Palladium and rhodium power catalytic converters. Silver is critical for solar photovoltaics and high-end electronics. Gold ensures reliability in connectors and advanced semiconductor packaging. These elements operate at the interface of clean energy, industrial chemistry, and digital systems, illustrating how the periodic table clusters economic importance around certain high-performance traits.

A different region of the map highlights the structural metals that make modern engineering possible. Titanium enables lightweight aerospace frames. Niobium strengthens superalloys used in jet engines. Tungsten provides density and hardness for cutting tools, armor, and missile systems. Hafnium and zirconium support nuclear reactors. Magnesium and aluminum enable lightweight transportation systems and components. These materials form the physical architecture of mobility, defense, and heavy industry. While some of them do not appear on every criticality list, their importance to aerospace and military sectors makes them strategic in practice if not in name.

Another cluster includes energy-transition materials that enable renewable power and electrification. Lithium, nickel, cobalt, graphite, manganese, and phosphorous form the chemical foundation of lithium-ion batteries. Copper—the oldest metal used by humans—has re-emerged as a cornerstone of the future, required in enormous quantities for grids, electric vehicles, motors, and renewable power systems. Aluminum supports transmission lines, photovoltaic frames, and numerous industrial applications. These materials represent the high-volume, high-growth segment of the critical materials landscape, where scarcity emerges not from geological rarity but from the speed of demand growth relative to the pace of supply expansion.

If we map the periodic table of power with geopolitical overlays, patterns become even more pronounced. Rare earths, gallium, germanium, and graphite are heavily concentrated in China, giving it dominant influence over clusters of the map associated with magnets, semiconductors, and batteries. Cobalt clusters map largely to the Democratic Republic of Congo, while nickel clusters map to Indonesia. Platinum group metals cluster around South Africa and Russia. Lithium clusters appear in Chile, Argentina, Bolivia, and Australia—with processing concentrated elsewhere. Copper clusters gather around Chile and Peru. Uranium clusters appear in Kazakhstan and Canada. These clusters illustrate why supply risk now shapes geopolitical strategy: critical materials are rarely distributed according to geopolitical preference.

The periodic table of power also highlights another important dimension: co-production and byproduct relationships. Many materials do not exist in isolation from an industrial perspective. Germanium is often a byproduct of zinc refining. Gallium emerges from bauxite processing. Indium comes from zinc and lead smelting. Tellurium and selenium are byproducts of copper refining. Dysprosium and terbium appear only in certain rare earth deposits. Nickel and cobalt are mined together in many contexts, especially in laterite deposits. These relationships complicate the supply landscape, as the availability of one material may depend on demand for another. For example, if zinc output falls, germanium supply may tighten even if demand remains constant. If copper refineries slow production, tellurium or selenium availability may be affected. These byproduct dynamics make the periodic table of power not just a static chart, but a network.

Another way to map power across the elements is to examine which technologies rely on which clusters. Digital communication depends on rare earths (erbium, yttrium), semiconductors (gallium, germanium, indium), and precious metals (gold). Energy storage depends on lithium, nickel, cobalt, graphite, and manganese. Renewable energy depends on silver, silicon, rare earth magnets, and copper. Aerospace depends on titanium, nickel, niobium, tungsten, and aluminum. Defense depends on rare earths, titanium, tungsten, beryllium, zirconium, hafnium, and several specialized alloys. AI and cloud computing depend on silicon, gallium, germanium, gold, palladium, copper, and high-performance ceramics.

By mapping these relationships, we see that power in the twenty-first century stems not from access to any single material but from mastery over combinations of materials. These combinations differ by sector. A country strong in copper, lithium, and nickel may dominate battery production but remain vulnerable in semiconductors. A country strong in rare earths may lead in magnets but remain exposed in platinum group metals. A country strong in gallium and germanium may shape semiconductor supply chains but rely heavily on other nations for battery materials or structural metals. No country controls the full map, which is precisely why materials have become a domain of intense strategic competition.

Mapping the periodic table of power also highlights differences between high-volume and high-leverage materials. Lithium, copper, and nickel have large, well-established markets measured in millions of tonnes. Their criticality arises from the sheer scale of demand growth and the difficulty of expanding supply quickly. By contrast, materials like gallium, germanium, hafnium, or dysprosium are used in far smaller quantities—sometimes measured in tonnes or even kilograms—but their leverage is much greater because they enable technologies that cannot function without them. In these cases, even small disruptions can have outsized consequences. This dynamic makes high-leverage materials disproportionately important in managing global risk.

Environmental and energy dimensions also shape the map. Mining rare earths, refining aluminum, processing graphite, and producing battery-grade chemicals consume significant energy. If energy constraints or environmental regulations affect production in key regions, the availability of materials shifts. This feedback loop—where energy security influences materials security, which in turn influences energy technologies—links the periodic table to the broader global system. A magnesium shortage caused by energy rationing affects aluminum alloys, which affects automotive supply chains. High energy prices reduce output of zinc smelters, indirectly tightening germanium supply. Mapping these relationships reveals a complex, interconnected system where materials, energy, and geopolitics form a triangle.

Innovation reshapes the map as well. Advances in substitution, recycling, and material science can shift the boundaries of criticality. New battery chemistries like lithium iron phosphate reduce the importance of cobalt and nickel. Advances in magnet design reduce dysprosium content. Quantum dots may reduce reliance on europium and terbium in certain optical applications. Sodium-ion batteries, if successful, could lessen dependence on lithium for some applications. But rarely do these innovations eliminate criticality altogether; they simply shift it toward new materials, such as higher purity manganese or enhanced anode materials. The periodic table of power therefore evolves with technology, not away from it.

Criticality is not an abstract label but an institutional judgment shaped by risk, dependence, and strategic exposure. Governments and organizations translate the material constraints of the periodic table into frameworks, lists, and policy priorities. As demand accelerates, these classifications become tools that influence investment, alliances, and the structure of the global system.

CHAPTER 8 – HOW CRITICALITY IS ASSESSED BY MAJOR INSTITUTIONS

Determining whether a raw material is "critical" is not a matter of intuition or geopolitical mood. It is the outcome of structured, quantitative, and regularly updated assessments conducted by major institutions around the world. These assessments combine economic analysis, supply-chain mapping, risk modeling, and sector-specific evaluations to identify materials that are essential to modern economies yet vulnerable to disruption. Although each institution adopts its own methodology and policy emphasis, they all share a similar conceptual core: criticality emerges at the intersection of high importance and high risk. Understanding these approaches is fundamental to understanding global materials strategy, especially as technologies evolve and geopolitical dynamics shift.

Criticality assessments start from a basic premise: not all materials are equally significant to the functioning of an economy or the resilience of its industrial systems. A material becomes critical when its absence would cause substantial economic damage or undermine national security, and when its supply is uncertain due to concentration, geopolitical influence, or limited substitutability. This dual structure—importance versus risk—appears across all major institutional frameworks.

The European Union provides one of the most systematic and transparent methodologies. Its criticality assessment, updated every three years, evaluates non-fuel, non-food raw materials using two main indicators: Economic Importance and Supply Risk. Economic Importance measures how essential a material is to the EU's industrial ecosystem, identifying the sectors that rely on it and estimating their contribution to the economy's value added. This value is then adjusted using a substitution index that reflects the difficulty of replacing the material with alternatives. A material that is indispensable in multiple high-value sectors, and for which substitutes are poor or non-existent, receives a high score.

Supply Risk, the second pillar of the EU's assessment, captures the vulnerability inherent in global supply chains. The EU analyzes how production is concentrated geographically, assesses the governance quality of supplier countries, and measures the EU's import reliance for that material. Critically, the EU evaluates risk not only at the mining stage but at whichever stage of the supply chain represents the bottleneck—an important nuance in a world where processing and refining are often more concentrated than extraction. Materials that exceed thresholds on both Economic Importance and Supply Risk enter the EU's list of Critical Raw Materials. In its 2023 update, the EU identified 34 CRMs from a candidate pool of 70, and introduced a new subgroup: Strategic Raw Materials. These are materials not only critical but essential to achieving green and digital transitions, and for which the EU has set explicit domestic supply targets under the Critical Raw Materials Act. This category includes lithium, cobalt, rare earths, nickel, and others whose disruption would significantly derail Europe's energy and industrial goals.

The United States uses a comparable but distinct approach. Under the Energy Act of 2020, a critical mineral is defined as a non-fuel resource essential to the economic or national security of the country and whose supply chain is vulnerable to disruption. The U.S. Geological Survey leads the assessment process, incorporating metrics such as production concentration, trade exposure, and the Herfindahl-Hirschman Index to quantify concentration risk. In addition, U.S. assessments emphasize the presence or absence of viable substitutes in manufacturing, which has proven crucial for materials like gallium, rare earths, and high-purity graphite. A 2021 USGS methodology paper introduced indices for economic vulnerability and disruption potential, defining thresholds that determine whether a material is included on the critical minerals list.

The U.S. list is updated periodically, reflecting technological and market developments. In 2022, the list expanded to 50 minerals, up from 35 in 2018. This expansion reflects both a more granular classification—such as treating individual rare earth elements separately—and evolving understandings of supply-chain fragility. Nickel, for example, was added due to its growing importance in battery cathodes and its concentrated production in Indonesia and Russia. Conversely, helium, potash, rhenium, and strontium were removed because updated assessments showed reduced supply risk or limited strategic importance. The U.S. process highlights a key reality: criticality is dynamic, not fixed.

Alongside the USGS process, the U.S. Department of Energy conducts its own Critical Materials Assessment focused specifically on energy technologies. This assessment is forward-looking, projecting demand under various clean energy scenarios and evaluating the likelihood of future supply–demand imbalances. Materials that may not appear critical in the present—such as lithium or manganese—may be identified as medium- or high-risk in DOE outlooks if projected demand growth vastly outpaces planned supply. In its 2023 assessment, the DOE flagged lithium, cobalt, nickel, and rare earths as the most critical materials for clean-energy technologies over the next decade. This approach underscores that future-oriented criticality can differ substantially from present-day criticality.

The International Energy Agency adopts yet another perspective. Rather than publishing a formal list, the IEA identifies minerals essential to achieving global climate and energy-security goals. In its influential reports on the role of minerals in clean energy transitions, the IEA highlights materials such as lithium, nickel, cobalt, copper, rare earths, manganese, graphite, and platinum-group metals as central to electric vehicles, wind turbines, solar PV, grid storage, hydrogen production, and related technologies. The IEA's approach is rooted in energy security: a mineral is effectively critical if its shortage would hinder the deployment of clean energy. This energy lens has become increasingly important as the pace of electrification accelerates. The agency also maintains a global policy tracker monitoring how governments respond to material vulnerabilities, reflecting the growing policy significance of critical minerals.

The World Bank brings yet another dimension: a development and climate perspective. In its analyses of mineral demand under climate scenarios, it identifies a set of materials that must scale dramatically—sometimes by hundreds of percent—to meet global decarbonization goals. These include lithium, cobalt, nickel, graphite, indium, vanadium, and several others. The World Bank emphasizes that the challenge is not scarcity in the Earth's crust but the ability of supply chains—especially in developing countries—to expand responsibly and rapidly enough to match clean energy demand. Although the World Bank does not use a strict supply-risk formula, its scenario-based assessments implicitly define criticality as the combination of high climate relevance and strong projected demand growth.

Across these institutions, several themes emerge. First, all assessments recognize the dual dimension of importance and risk. A material can be economically vital yet secure, or risky yet not particularly important; only when both conditions coincide does criticality emerge. Second, the assessments acknowledge that criticality is fluid. Technology evolves, demand patterns shift, geopolitical dynamics change, and what was non-critical yesterday may become crucial tomorrow. Third, criticality is context-dependent. The EU may classify phosphate rock as critical due to fertilizer dependence, while the U.S. may not. The U.S. may consider uranium critical due to national security concerns, while the EU applies different criteria. The IEA focuses on energy-transition minerals, while the World Bank focuses on minerals essential to climate mitigation and development.

Another insight from these frameworks is that criticality increasingly depends on processing, not just extraction. The EU explicitly evaluates risk at the narrowest bottleneck in the chain. The U.S. highlights that a material can be mined widely but refined narrowly, creating supply risk even where geological abundance is high. The IEA's emphasis on concentration in refining for lithium, cobalt, and rare earths further reinforces this point. This shift reflects the reality that supply chains are global, multi-stage, and often dominated by countries that specialize in processing rather than extraction.

Criticality assessments also highlight the role of substitution and recycling. Where substitution is easy or effective, criticality declines. Where substitution is difficult—such as for rare earth magnets or gallium-based semiconductors—criticality increases. Recycling potential can mitigate long-term supply risk, although short-term constraints remain due to low recovery rates and limited recycling infrastructure. The ability to recycle cobalt or recover rare earths from magnets, for instance, plays a growing role in institutional assessments, particularly in the EU and U.S., which emphasize circular-economy strategies.

CHAPTER 9 – CLASSIFICATIONS: US, EU, IEA AND GLOBAL FRAMEWORKS

As critical raw materials have become central to industrial strategy and geopolitics, different institutions and governments have developed their own classification systems to identify which materials deserve priority. These frameworks are more than academic exercises: they guide billions in investment, shape regulatory decisions, determine stockpiling strategies, influence trade policy, and set the direction for energy and innovation planning. Understanding the similarities and differences among the U.S., European Union, International Energy Agency, and other global frameworks is therefore essential for grasping how nations interpret and respond to material risks.

Although each classification system reflects its own political, industrial, and economic context, they are all built on a similar conceptual foundation: a critical material is one that is economically essential and at risk of supply disruption. Yet the criteria, thresholds, and policy implications differ significantly across regions. These differences reveal not only the diversity of industrial structures and political priorities but also the varying degrees of vulnerability among global actors.

The European Union's approach is one of the most structured and data-driven. Its Critical Raw Materials list, updated every three years, identifies materials that meet quantitative thresholds for economic importance and supply risk. Beginning with a large candidate pool, the EU assesses each material's role in key industrial ecosystems—such as aerospace, automotive, electronics, renewable energy, and digital sectors—and evaluates whether substitutes are technologically or economically feasible. It then assesses supply risk based on import dependence, production concentration, and the governance strength of supplier countries. Materials that score highly in both dimensions are included on the list.

The EU's classification system is also unique in that it introduces a second tier: Strategic Raw Materials. These are the materials considered indispensable for the EU's green and digital transitions and for strategic technologies such as batteries, renewable energy systems, semiconductors, robotics, and space applications. Strategic materials include lithium, cobalt, nickel, rare earth elements, silicon metal, gallium, and others. Under the Critical Raw Materials Act, the EU has set ambitious targets for domestic extraction, processing, refining, and recycling of strategic materials. This dual-structured classification—critical versus strategic—reflects Europe's desire not only to identify materials at risk but to direct industrial policy toward securing long-term sovereignty.

The United States adopts a different structure. The U.S. Geological Survey's Critical Minerals List, updated periodically, identifies minerals essential to economic and national security and vulnerable to supply disruption. Unlike the EU system, which uses two specific indices, the U.S. methodology uses a more flexible combination of metrics: production concentration, global market dynamics, import reliance, disruption potential, substitutability, and the strategic contribution of each material to defense and high-tech industries. The U.S. list expanded from 35 materials in 2018 to 50 in 2022, reflecting a more granular understanding of supply-chain dependence. For example, the U.S. distinguishes among individual rare earth elements rather than treating them as a single group, and it includes materials like nickel or zinc when their supply chains become increasingly vulnerable.

The U.S. system also incorporates national security considerations more explicitly than the EU framework. Materials used in defense—tungsten, titanium, rare earths, beryllium, hafnium, and others—are prioritized because supply interruptions would compromise military readiness. Additionally, U.S. classifications feed directly into policy through mechanisms such as the Defense Production Act, strategic stockpiles, critical minerals grants, and investment restrictions on foreign entities. The Department of Energy's own Critical Materials Assessment adds another layer by focusing specifically on energy technologies and projecting future supply-demand imbalances under different climate and electrification scenarios. This makes the U.S. approach both multi-institutional and forward-looking, reflecting the country's reliance on advanced technologies and its desire to reduce dependence on geopolitical competitors.

The International Energy Agency uses yet another approach, oriented around the global energy transition. Unlike the EU and U.S., the IEA does not publish a formal list of "critical minerals." Instead, it identifies minerals essential to clean energy technologies and assesses the vulnerability of their supply chains. The IEA focuses on materials used in electric vehicles, solar PV, wind turbines, hydrogen electrolyzers, and grid technologies. These include lithium, cobalt, nickel, rare earths, copper, manganese, graphite, silicon, silver, platinum-group metals, and others. Its analysis emphasizes three themes: the role of minerals in clean energy technology, the pace at which supply needs to expand to meet climate goals, and the geopolitical implications of concentrated refining and processing.

The IEA's framework is particularly notable for highlighting how different clean energy pathways influence which materials become critical. For example, aggressive EV adoption increases demand for lithium and nickel; rapid solar expansion increases demand for silver and silicon; and widespread hydrogen deployment raises the importance of platinum and iridium. In this way, the IEA situates criticality within the broader context of global climate policy and energy security. Rather than assigning a binary classification—critical or not—the IEA analyzes materials along a spectrum of vulnerability depending on technology mix, market concentration, and expected demand curves.

Global organizations outside the U.S., EU, and IEA also contribute to classification frameworks. The World Bank's assessments emphasize materials central to climate mitigation and sustainable development, identifying those that must scale dramatically to support clean energy in emerging economies. Although not a formal criticality list, the World Bank highlights materials such as lithium, cobalt, nickel, copper, molybdenum, and rare earths as essential to decarbonization pathways, focusing on responsible extraction and development implications. Japan, South Korea, and Australia maintain their own lists as well, reflecting regional industrial structures. Japan emphasizes materials essential to its advanced manufacturing and electronics industries, such as gallium, germanium, rare earths, and tungsten. South Korea focuses heavily on battery materials, reflecting its dominant EV and storage sectors. Australia, as a major producer of lithium and rare earths, focuses on securing downstream processing capability and diversifying export partnerships.

Comparing these frameworks reveals several patterns. First, there is substantial overlap among the materials classified as critical across institutions. Rare earth elements consistently appear on nearly every list. So do lithium, cobalt, nickel, gallium, germanium, graphite, and platinum-group metals. Copper, while not always formally included on every critical list, features prominently in IEA and World Bank assessments due to its essential role in electrification. This convergence reflects the underlying reality: modern technology systems around the world depend on the same foundational materials.

Yet important differences also emerge. The EU includes materials like bauxite, phosphate rock, and magnesium due to their centrality in European industrial supply chains, whereas the U.S. excludes some of these but includes others such as helium or barite. Russia's role in palladium and platinum influences some national assessments but not others. Japan's reliance on imported resources leads to a more expansive classification. Australia, as a resource-rich nation, tends to prioritize materials where it seeks to expand processing and reduce global dependence on China. These differences reveal that criticality is not a universal category but a lens shaped by each region's industrial structure, geopolitical position, and strategic ambitions.

Another distinction is how institutions interpret supply-chain vulnerability. The EU tends to prioritize governance risk—evaluating how political stability, transparency, and regulatory quality in supplier countries affect reliability. The U.S. gives more weight to national security exposure and the strategic consequences of dependence on adversarial nations. The IEA measures vulnerability through the lens of the energy transition, emphasizing bottlenecks in refining and the pace at which supply must expand. The World Bank considers environmental and social constraints that influence long-term availability. These distinct emphases reflect institutional missions but also highlight the many dimensions of supply risk—political, economic, technological, environmental, and temporal.

A further difference lies in how institutions incorporate the concept of substitution. The EU explicitly quantifies it; the U.S. incorporates it qualitatively; and the IEA analyzes substitution primarily in the context of technology pathways. Materials like cobalt, which face substitution pressure from new battery chemistries, may appear critical in some frameworks but less so in others. This underscores the dynamic nature of technology and its ability to shift material requirements on relatively short time horizons.

Despite these differences, global frameworks are converging in important ways. All major institutions recognize the central role of processing and refining bottlenecks, the urgency of diversifying supply chains, and the need to scale production sustainably. All acknowledge that demand for many materials needed for clean energy and advanced electronics will grow faster than supply can easily adjust. All emphasize the importance of recycling, substitution research, and collaborative international approaches to securing long-term material resilience. This convergence creates the foundation for cross-border coordination, even as national strategies diverge in implementation.

Classifications shape policy responses. In the EU, they activate support for domestic projects and influence foreign partnerships. In the U.S., they trigger defense procurement mechanisms and trade interventions. In the IEA's work, they guide energy security and investment recommendations. In all cases, these frameworks serve as strategic maps—identifying vulnerabilities, guiding innovation, and shaping how governments and industries prepare for a more resource-intensive technological future.

Critical materials only acquire strategic meaning when embedded in industrial systems and supply chains. It is there that classification turns into leverage, vulnerability, and power. The movement of these materials through extraction, processing, and manufacturing ultimately determines who captures value and who bears risk in the global order.

PART II – SUPPLY CHAINS: MINING, REFINING, PROCESSING

CHAPTER 10 – FROM ORE TO OXIDE: EXTRACTING RARE EARTHS

The journey from rare earth ore to purified oxide is one of the most complex, technically demanding, and environmentally sensitive processes in the entire critical materials landscape. Unlike many metals that can be extracted and refined through relatively straightforward steps, rare earth elements are interlocked chemically, dispersed in diverse geological formations, and require elaborate multi-stage processing before they become usable in high-tech applications. Understanding this transformation—from rock to concentrate, from concentrate to solution, and from solution to oxide—reveals why rare earth supply chains have become such powerful strategic levers and why only a handful of countries have mastered each stage.

The first link in the chain is geological. Rare earths are not found as pure metals but locked within specific minerals. The most commercially significant deposits fall into three categories: carbonatites rich in bastnäsite, heavy-element–rich ion-adsorption clays, and monazite-bearing sands. Each type demands different extraction approaches. Carbonatite deposits such as Bayan Obo in China and Mountain Pass in California contain bastnäsite, a fluorocarbonate mineral that can be mined through traditional open-pit operations. Monazite—a phosphate mineral—is often found in heavy mineral sands historically mined in Australia, India, and Brazil. Ion-adsorption clays in southern China and Myanmar represent another class, containing rare earths loosely bound to clay particles rather than embedded in crystalline structures. These geological differences shape everything that happens afterward.

Mining rare earths from hard rock deposits begins with drilling, blasting, and hauling. Ore grades are typically low, meaning large volumes of rock must be moved and milled to liberate the minerals of interest. After extraction, the ore is crushed and ground into fine particles, then subjected to physical separation procedures designed to produce a concentrated rare earth mineral product. Flotation is the main technique for bastnäsite and monazite ores: reagents cause rare earth minerals to attach to bubbles and rise in flotation cells, separating them from surrounding rock. Magnetic separation and gravity separation sometimes complement flotation, depending on the precise mineralogy. The goal is to upgrade the raw ore into a concentrate containing perhaps 50–70 percent rare earth minerals. Only then can chemical processing begin.

Ion-adsorption clays follow a radically different path. Instead of blasting rock, operators drill shallow wells into the clay layers and pump in leaching solutions, typically ammonium sulfate or similar salts. The rare earth ions—weakly adhered to the clay matrix—exchange with ammonium ions, dissolve into solution, and are pumped back to the surface as a "pregnant leach solution." This method requires no crushing or milling and leaves no tailings piles, but it presents its own risks: if poorly managed, the leaching fluids can contaminate soil and groundwater. The scalability and environmental impact of ion-adsorption clay mining have become major points of global interest because these clays contain some of the world's richest natural sources of heavy rare earth elements.

Once a mineral concentrate or pregnant leach solution has been produced, the material moves to the next phase: cracking, leaching, and dissolution. For bastnäsite and monazite, the concentrate must typically be roasted or chemically cracked to break down the mineral structure and render the rare earths soluble. Bastnäsite concentrates are often roasted in the presence of air or chemical additives to remove carbonates and fluorides, converting the material into oxides or oxyfluorides that can later be dissolved. Monazite cracking frequently involves strong acids, such as sulfuric or hydrochloric acid, to break apart the phosphate structure. The process is energy- and reagent-intensive and can generate radioactive residues because monazite often contains thorium. Managing these residues safely has become a defining challenge for the industry outside China, where historical environmental laxity helped establish processing dominance.

After cracking, the material undergoes leaching, which dissolves the rare earths into solution. Whether the leach medium is acidic or alkaline depends on the ore chemistry and the downstream processing requirements. The goal is to produce a clear, metal-rich solution containing a cocktail of rare earth ions—lanthanum, cerium, neodymium, praseodymium, and a host of others—combined in a single, undifferentiated liquid. This marks the point at which the true complexity of rare earth processing becomes apparent.

The heart of rare earth refinement is separation. Although rare earths have different atomic numbers, their chemical behavior is notoriously similar. Each element has nearly the same ionic radius and valence, making the task of separating them exceptionally difficult. The solution to this problem, developed over decades, is solvent extraction: a series of hundreds of liquid–liquid separation steps that progressively isolate each element. In a solvent extraction plant, the leach solution passes through long arrays of mixer-settler units where it contacts organic solvents that selectively bind or repel specific rare earth ions. Slight differences in solubility are amplified across hundreds of cycles until each element accumulates in its own stream. This process requires vast chemical volumes, precise controls, and highly trained operators.

The scale and sophistication of solvent extraction plants explain why separation capacity has become the major bottleneck in the rare earth supply chain. Even where mining has diversified, most global separation still occurs in China, which built enormous capacity during years when environmental standards were lenient. China's head start allowed its processors to perfect the chemistry, enabling them to offer lower costs and unmatched throughput. As a result, even mines operating elsewhere often ship their concentrate to China for separation. The world's reliance on a single region for this critical step mirrors the dependence seen in other materials—nickel sulfates, cobalt hydroxides, graphite purification—but the rarity of separation expertise makes rare earths an especially stark case.

From separated streams, each rare earth is precipitated as a salt—often as an oxalate—and then calcined at high temperatures to create the final rare earth oxide. This is the primary form in which rare earths enter global markets: neodymium oxide, praseodymium oxide, dysprosium oxide, europium oxide, and others. These oxides are powders of varying purity and composition, but they are still several steps away from becoming functional materials. Converting oxides into metals or alloys requires high-temperature reduction processes, such as metallothermic reduction using calcium or electrolytic methods for certain elements. Metal-making has historically been less concentrated than separation, but China also dominates much of this stage. The resulting metals are then alloyed into magnet alloys, phosphor compounds, catalysts, optical materials, and other specialized feedstocks.

Environmental management is a defining challenge at every stage. Rare earth mining can generate large tailings ponds containing radioactive or chemically hazardous waste. Ion-adsorption clay extraction, if mismanaged, can contaminate groundwater. Cracking and leaching require concentrated acids and produce waste streams that must be neutralized and contained. Solvent extraction plants use large volumes of organic solvents that must be carefully recycled or destroyed. The global push for cleaner supply chains has forced new entrants to adopt far stricter environmental controls than those used during the period when Chinese dominance was established. This raises capital costs but is indispensable for developing a sustainable rare earth industry.

Energy consumption adds another layer of complexity. Producing one tonne of rare earth oxide requires significant amounts of electricity, chemical reagents, heat, and water. These requirements vary by deposit and processing route, but the general pattern is clear: rare earth processing is far more energy-intensive than most people imagine. As countries attempt to build more localized supply chains, ensuring access to low-carbon, low-cost energy becomes a competitive advantage. In some regions, renewable energy or hydropower can reduce the carbon footprint of processing. In others, dependence on fossil-intensive grids could undermine the environmental benefits of downstream technologies like electric vehicles and wind turbines.

Despite the challenges, the world is seeing renewed interest in expanding rare earth extraction and separation capacity. Projects in Australia, the United States, Canada, Europe, Vietnam, and parts of Africa aim to replicate capabilities that have long been concentrated in China. New technologies are also emerging. Improved solvent extraction chemistries, membrane-based separation, ion-exchange resins, and even bioleaching approaches are being tested. Some companies are exploring recycling rare earths from end-of-life magnets and electronics, a pathway that could eventually offset primary extraction demand. While recycling volumes remain small today, the long-term potential is significant as EV motors, wind turbines, and consumer devices reach end of life in larger quantities.

The transformation from ore to oxide is therefore much more than a technical process. It is a geopolitical journey, a test of industrial capacity, and an environmental balancing act. Rare earth processing exemplifies how modern technologies rely not only on scientific innovation but on intricate supply chains that must be carefully managed, protected, and expanded. The next chapter will examine the global geography of mining—where rare earths and other critical materials are extracted, who leads, who lags, and how geology and politics intertwine to shape the upstream foundations of the energy transition.

CHAPTER 11 – GLOBAL MINING GEOGRAPHIES: WINNERS AND LAGGARDS

The global geography of mining critical raw materials reveals an asymmetric landscape shaped by geology, history, politics, and industrial strategy. These imbalances are not accidental. They reflect how nature distributed mineral deposits millions of years ago, but also how nations invested—or failed to invest—in extraction, refining, infrastructure, and environmental governance. The result is a world where a handful of countries dominate the upstream segment of essential supply chains, while others remain dependent, vulnerable, or strategically lagging. Mapping this geography is crucial for understanding global power dynamics and for anticipating future supply disruptions.

The most striking feature of today's mining landscape is concentration. A small number of countries supply most of the world's critical materials. For some minerals, production is not merely concentrated—it is effectively monopolized. The Democratic Republic of Congo produces nearly three quarters of the world's cobalt, a figure that underscores the country's unique geological endowment but also reflects the limited number of nations with comparable resources. China produces the majority of rare earth elements and an overwhelming share of heavy rare earths. Indonesia dominates nickel production and controls a large share of global reserves. Chile is the world leader in lithium reserves and a major producer, while Australia dominates lithium mining volumes. South Africa leads in platinum group metals, and Kazakhstan remains a key player in uranium. These geographic concentrations shape supply chains, prices, and geopolitical strategies in ways few other commodities do.

Rare earth mining is one of the clearest examples of geographic dominance. China's Bayan Obo deposit in Inner Mongolia is the world's single largest source of rare earths. Although new mines have emerged in the United States and Australia, China continues to supply a majority of global rare earth oxides and remains the only significant producer of heavy elements such as dysprosium and terbium. Myanmar has recently become an important contributor, particularly through ion-adsorption clay deposits, but its production is unstable, heavily influenced by local conflict, informal operations, and Chinese involvement. This dynamic reinforces China's central role, as much of Myanmar's output eventually flows into Chinese separation facilities. Other countries, including Vietnam, Brazil, and India, possess promising deposits, yet challenges in permitting, capital availability, and environmental regulation limit their ability to scale quickly.

For cobalt, the Democratic Republic of Congo stands unmatched. The copper belt that stretches across southern DRC and northern Zambia hosts some of the world's richest cobalt-bearing deposits. Operations such as Tenke Fungurume, Kamoa-Kakula, and Mutanda supply the majority of global cobalt used in lithium-ion batteries. Much of this production is linked to Chinese companies, who acquired assets during a period of aggressive overseas investment. This has created a dual concentration, both geographic and corporate. Efforts from the United States, Europe, and Japan to diversify cobalt supply have led to investments in Australia, Canada, and emerging projects in Indonesia, but these will take time to meaningfully shift the balance.

Nickel mining also reveals stark disparities. Indonesia has rapidly evolved from a major ore exporter to the world's largest nickel producer after banning raw ore exports and forcing companies to invest in domestic processing. Its strategy—intended to capture more value domestically—has drawn significant foreign investment, particularly from China, and altered global supply chains. The Philippines remains a significant contributor, though its ore quality varies, and its environmental policies have been inconsistent. Russia, Canada, and Australia supply additional volumes, though production in Russia faces ongoing geopolitical uncertainty and potential sanctions that could reshape global nickel markets. As demand for high-purity nickel sulfate increases due to electric vehicle batteries, the geography of refined nickel rather than mined nickel becomes even more strategic.

Lithium mining is somewhat more geographically distributed, yet still dominated by a few countries. Australia produces over half of global lithium, primarily from hard-rock spodumene deposits concentrated in Western Australia. Chile and Argentina form the core of the South American "lithium triangle," home to vast brine resources in high-altitude salt flats. Bolivia holds immense reserves but has struggled for decades to develop them due to technological, political, and economic barriers. China contributes additional supply from both hard-rock and brine deposits, but its global role is even more pronounced in refining rather than mining. New producers like Canada, Brazil, and Zimbabwe are entering the market, but none yet rival the established giants.

Copper mining represents another central pillar of the energy transition. Chile is the world's leading producer, followed by Peru. These two countries together supply nearly 40 percent of global copper. Yet their mining sectors face challenges: declining ore grades, water scarcity intensified by climate change, political tensions over royalties, and increasing social resistance to new projects. These pressures could slow future output just as electrification surges. Other key producers include China, the United States, the Democratic Republic of Congo, Zambia, and Indonesia. Mongolia's Oyu Tolgoi mine and Panama's Cobre Panamá project (before its political shutdown) represent major additions to global supply, though both highlight the growing friction between mining interests and national politics.

Platinum group metals remain concentrated in Southern Africa. South Africa dominates platinum and rhodium production, while Russia provides a major share of palladium. The fragility of this supply became evident during geopolitical events that constrained Russian exports. South Africa's mining operations, meanwhile, face electricity shortages, labor issues, and aging infrastructure, adding operational risk to the global PGM market. These materials are essential to catalytic converters, hydrogen technologies, and many industrial applications, making their geographic concentration a persistent concern for manufacturers.

Graphite presents yet another example of concentration. China produces the vast majority of natural graphite and nearly all battery-grade purified graphite. Mozambique has recently emerged as a significant producer, with major projects supplying natural graphite to global markets. Nevertheless, raw graphite is only part of the story: the purification process is heavily dominated by China, meaning that even graphite mined elsewhere often flows through Chinese refineries before reaching battery manufacturers.

Uranium production is similarly skewed. Kazakhstan leads global uranium mining, supported by low-cost in-situ recovery operations. Canada and Australia follow, but the prominence of Kazakhstan creates dependency challenges for countries seeking to deploy or maintain nuclear power. Political dynamics and security concerns in Central Asia further complicate long-term stability of supply.

Some countries with substantial mineral endowments remain underperformers. India, for example, holds significant monazite-bearing sand deposits and rare earth potential but has struggled to scale production due to regulatory barriers and bureaucratic complexity. Similar patterns appear in Brazil, which possesses both rare earth and niobium resources, and in African nations with abundant geological potential but limited infrastructure, inconsistent governance, or conflicting regulatory frameworks. These "laggards" are not necessarily resource-poor; instead, they face structural constraints that inhibit investment and development. Their future role will depend on improvements in governance, environmental management, and partnerships that balance local interests with global demand.

Environmental and social factors further influence mining geographies. In countries with strong environmental regulations, such as those in Europe or North America, new mining projects face lengthy permitting timelines and public opposition. Debates over land rights, water usage, biodiversity, and indigenous consultation can significantly delay or block developments. While these concerns promote responsible mining, they also restrict diversification away from regions with weaker regulations. This dynamic has limited new rare earth, lithium, and nickel projects in Western nations and slowed progress on critical minerals needed for the energy transition.

Several regions have emerged as new frontiers. Africa is increasingly seen as a key partner in future supply chains—not only for cobalt, but also for lithium, graphite, manganese, rare earths, and copper. Zimbabwe and Namibia have growing lithium sectors, while Malawi, Tanzania, and Burundi hold rare earth deposits that could challenge global concentration if developed responsibly. Latin America remains central to lithium but also offers potential expansion in copper, nickel, and rare earths. Canada presents opportunities across a broad suite of materials, supported by strong regulatory frameworks and growing alliances with the United States and Europe. Even the Middle East, historically synonymous with oil, has begun exploring opportunities in critical minerals, particularly through investments in Africa and Asia.

Mining geographies also intersect with geopolitical strategies. China's overseas investments in Africa, South America, and Southeast Asia have secured access to cobalt, copper, nickel, lithium, and bauxite, reinforcing its role as a global processing hub. Western countries are racing to catch up through partnerships, development finance, and diplomatic agreements. Initiatives such as the Minerals Security Partnership aim to diversify supply chains and encourage responsible mining in emerging markets. These efforts highlight that geography alone does not determine influence; investment, infrastructure, and processing capacity often matter just as much as ore bodies.

Climate change adds another layer of complexity. Many of the world's top mining regions face increasing environmental stress. Water scarcity threatens lithium brine projects in the Andes and copper mines in Chile. Extreme weather disrupts operations in tropical regions. Rising temperatures affect workers and energy systems. These factors could reduce output or require costly mitigation measures, shifting the attractiveness of different regions for future mining investment.

Refining and chemical processing are where material flows narrow and strategic exposure sharpens. At this stage, technical complexity and capital intensity compress production into a limited number of locations. Control over these chokepoints often matters more than access to the raw resource itself.

CHAPTER 12 – REFINING BOTTLENECKS AND CHEMICAL PROCESSING

Refining is the quiet center of the critical materials universe. Mining attracts attention because it is physically visible, politically contentious, and often environmentally disruptive. But it is the refining and chemical processing stage that truly determines who controls a supply chain. The transformation of ore into highly purified oxides, chemicals, metals, and intermediate feedstocks is where the majority of value is added and where the greatest bottlenecks exist. Unlike mining, which can occur in many countries with the right geology and sufficient capital, refining requires deep technical expertise, sustained investment, abundant energy, strict process control, and decades of accumulated know-how. These are the constraints that define the global landscape of critical materials—and explain why refining, far more than extraction, is the center of geopolitical leverage.

The refinement of critical materials is governed by chemistry. Each element has its own behaviors, solubilities, oxidation states, ionic radii, and reaction pathways that must be managed with precision. Ore bodies vary in composition and impurities, requiring tailored processing lines that cannot easily be repurposed. Moreover, refining rarely produces a single product: most critical materials emerge as co-products or byproducts of more mature industries. Gallium is extracted from the alkaline liquor of alumina refineries. Germanium is recovered from zinc smelter residues or coal fly ash. Tellurium and selenium appear in copper refinery slimes. Cobalt often emerges from copper mines; nickel often appears alongside cobalt. This interdependence means that the availability of one material often depends on the economics and throughput of another industry entirely. A slowdown in zinc smelting affects germanium supply; a shift in copper markets affects tellurium; an unexpected closure of an alumina refinery can constrict gallium overnight.

Refining bottlenecks matter because they are the narrowest points in global supply chains. For many critical materials, mining is relatively diversified, but refining is heavily concentrated. The global economy now depends on just a handful of countries to perform the chemical steps required to convert raw minerals into usable materials. China dominates rare earth separation, lithium refining, cobalt and nickel chemicals, graphite purification, gallium and germanium recovery, and magnet alloy production. This dominance is not accidental. It was built through decades of investment, policies encouraging domestic value addition, low energy costs, loose environmental regulation during critical periods of scale-up, and strategic acquisitions of overseas assets. The result is an international system where even if extraction diversifies, chemical processing often remains tied to China.

Rare earth refining illustrates the challenge vividly. Once a mixed concentrate or pregnant leach solution is produced, the crucial step is separation. Rare earth elements have exceptionally similar ionic radii and chemical behavior, making their separation one of the most technically complex tasks in modern metallurgy. Solvent extraction, the dominant method, requires hundreds of mixer-settler stages where slight differences in solubility and binding affinity are amplified until each element is isolated in its own stream. This process consumes enormous quantities of organic solvents, acids, bases, water, and energy. Only a few countries have mastered it. China built its capacity during a period of regulatory permissiveness that would not be possible today in most democracies. Although new rare earth projects are being developed in the United States, Australia, Canada, and Europe, the separation stage remains a bottleneck that takes years—not months—to replicate.

Lithium refining presents its own complexities. Hard-rock spodumene must be roasted at high temperatures, converted into a leachable phase, dissolved with sulfuric acid, purified, and crystallized into lithium carbonate or hydroxide. Brine-derived lithium requires long evaporation cycles or emerging direct extraction technologies to yield a concentrated lithium chloride feedstock that must then be chemically processed into battery-grade compounds. Battery-grade lithium is not simply a commodity; it must meet strict purity thresholds because impurities can degrade battery performance and safety. Even slight variations in particle morphology or residual metals can disrupt cathode manufacturing. As a result, lithium refining is concentrated not just geographically but technologically, with a handful of companies supplying most of the world's battery-grade chemicals.

Cobalt and nickel refining also present structural bottlenecks. Cobalt hydroxide from the Democratic Republic of Congo must be refined into cobalt sulfate or metal, a process dominated by Chinese refiners. Nickel laterite ores, abundant in Indonesia and the Philippines, require energy-intensive processing routes such as high-pressure acid leaching, which uses autoclaves operating at extreme conditions. These facilities cost billions to build, face significant technical risks, and require stable power supplies. Attempts to build such refineries outside Asia have frequently encountered delays, cost overruns, and operational challenges. Nickel sulfide ores, processed through flotation and smelting, yield nickel matte that must then be refined into high-purity class 1 nickel. This final step, especially when producing battery-grade nickel sulfate, remains a technical barrier that only a limited number of companies and countries have overcome.

Graphite purification highlights the environmental dimension of refining bottlenecks. Natural graphite must be purified to more than 99.9 percent before it can be shaped into spherical particles for lithium-ion battery anodes. In China, this is typically achieved using hydrofluoric acid, a hazardous chemical that requires precise handling and creates disposal challenges. Outside China, companies are investing in thermal purification and alternative processes, but these methods require significant energy and are not yet cost-competitive at scale. As a result, even when graphite is mined in Africa or elsewhere, purification tends to occur in China, reinforcing dependence.

Gallium and germanium, which are essential for semiconductors, fiber-optics, infrared optics, and advanced electronics, depend on refining systems deeply embedded within base-metal industries. Because they are byproducts, their supply depends on smelter operating rates. When global aluminum output falls, gallium recovery drops; when zinc smelters cut production due to high energy prices, germanium supply tightens. This dynamic was made clear in Europe when rising electricity costs forced smelter shutdowns, constricting the availability of minor metals and exposing how closely materials security is tied to energy security.

Energy itself is a hidden but critical component of refining. Many processes consume vast quantities of electricity, heat, or both. Aluminum smelters require enormous amounts of power; magnesium production, particularly in China, depends on coal-fired electricity. Nickel HPAL plants are among the most energy-intensive refining systems in the world. Rare earth separation circuits require continuous power to operate pumps, mixers, heaters, and controls across hundreds of stages. If a country has high energy prices, unstable grids, or carbon-intensive power, its refining competitiveness suffers. This is one reason energy shocks, such as those in Europe, can shut down metal production overnight and trigger shortages across global supply chains. It also means that decarbonizing the refining segment—by deploying renewable power, hydrogen, or advanced energy systems—is becoming essential not only for climate goals but for long-term competitiveness.

Environmental constraints shape refining just as profoundly as energy. The chemical processes involved generate waste streams that must be neutralized, treated, and disposed of safely. Rare earth separation produces large volumes of acidic waste and can leave behind radioactive residues if monazite is used. Lithium brine extraction risks depleting water in arid regions, while hard-rock refining can release sulfuric-acid-contaminated effluents. Nickel laterite processing produces massive waste ponds, and tailings management remains a persistent concern across multiple minerals. These barriers mean that new refining facilities face substantial regulatory hurdles in many countries, making it difficult to replicate capacity quickly.

Refining bottlenecks create fragility even in long supply chains. When a single refinery in China, Europe, or Southeast Asia shuts down due to power shortages, environmental enforcement, or maintenance, global downstream industries feel the impact. The industrial structure amplifies the risk: many companies keep minimal inventories due to just-in-time manufacturing practices. A disruption in a cobalt refinery or rare earth separation plant can quickly undermine battery manufacturing, automotive production, or wind turbine assembly. The 2010 rare earth incident demonstrated how quickly prices can skyrocket when a major producer restricts exports. More recently, export controls on gallium and germanium revealed how supply-chain chokepoints can be used as geopolitical tools, triggering shortages in semiconductor and optics industries.

The concentration of refining capacity creates a power asymmetry. Countries that mine critical materials must often ship them to overseas processors, losing value addition and exposing themselves to political risk. Even well-capitalized mining nations struggle to build refining industries because of the long learning curves, high capital costs, environmental constraints, and market volatility. This dynamic explains why China's position at the center of refining is so difficult to challenge: the infrastructure is not simply large; it is deeply integrated, technologically advanced, and linked to a vast ecosystem of downstream manufacturers.

Diversification efforts are underway, but progress is uneven. The United States, Europe, Japan, South Korea, Australia, and Canada are investing in refining capacity for rare earths, lithium, cobalt, nickel, and graphite. Several projects seek to commercialize cleaner refining technologies, such as direct lithium extraction, alternative graphite purification, rare earth magnet recycling, and solvent extraction innovations. Governments are offering subsidies, loan guarantees, permitting reform, and strategic partnerships to overcome initial barriers. Yet scaling these systems takes time, and the world remains years away from meaningful diversification in most materials.

Innovation will play an increasingly important role. New refining chemistries, more selective extractants, membrane-based separations, and bioleaching approaches may eventually reduce environmental impact and lower costs. Recycling will also become essential as end-of-life batteries, wind turbines, electronics, and motors enter the waste stream in large volumes. Over time, secondary supply could reduce pressure on mining and refining, but this shift will not occur at scale until the 2030s or beyond.

Refining and chemical processing are therefore the crucible in which global material power is forged. The world's dependence on a handful of refining hubs shapes geopolitical relationships, corporate strategies, and national policy decisions. Resilience is forged less at the mine than in the industrial systems that transform ore into high-purity inputs at scale. Mastery of processing, refinement, and downstream integration determines whether resources become strategic assets or persistent vulnerabilities. Where this capability concentrates, so does influence over the global material system.

CHAPTER 13 – CHINA'S DOMINANCE IN THE VALUE CHAIN

China's dominance in the critical materials value chain did not emerge suddenly or by accident. It is the result of decades of deliberate strategy, industrial policy, technological investment, and geopolitical foresight. Today, China occupies a commanding position across virtually every stage of supply chains for rare earths, lithium, cobalt, nickel, graphite, gallium, germanium, and a wide constellation of other strategic materials. This dominance extends beyond mining into the more decisive stages of separation, chemical refining, metal-making, component manufacturing, and the fabrication of end-use technologies. Understanding how China achieved this position—and what it means for the global economy—is essential for anyone seeking to comprehend the modern geopolitical landscape.

China's rise began with rare earths, a field in which it remains unrivaled. While countries such as the United States once led in mining and processing, China steadily expanded its capabilities through the 1980s and 1990s, leveraging lower production costs, fewer environmental constraints, and coordinated industrial policy. The country built a suite of state-backed companies that perfected the complex solvent-extraction technologies required to separate individual rare earth elements. Over time, China accumulated the operational expertise, supply-chain infrastructure, and industrial base that allowed it to dominate the global rare earth market. Even today, as mining gradually diversifies, the bottleneck remains in separation and metal-making—stages overwhelmingly controlled by China. Mines in the United States, Australia, and elsewhere frequently send their concentrates to Chinese facilities for processing, demonstrating how deeply integrated China has become in this segment of the chain.

China's strategy consistently linked upstream resources to downstream industrial power. This linkage allowed it to capture not just minerals, but also the high-value products made from them. Rare earth magnets illustrate this dynamic. China produces the vast majority of neodymium-iron-boron magnets used in electric vehicles, wind turbines, data centers, and advanced electronics. Its magnet manufacturers benefit from proximity to separated rare earth oxides, dense networks of component suppliers, and strong domestic demand from automotive, electronics, and renewable energy industries. The result is a level of vertical integration unmatched by any other country. Even when foreign companies attempt to build magnet factories, they often remain dependent on Chinese refiners or alloy producers.

Lithium refining is another pillar of China's dominance. While Australia leads in hard-rock lithium mining, most spodumene concentrate still flows into Chinese conversion facilities, where it is transformed into battery-grade lithium carbonate and hydroxide. China refined well over half of the world's lithium chemicals in recent years. Its refiners maintain an advantage through scale, decades of experience, integrated supply chains, and close ties to the world's leading battery manufacturers. As a result, nations that mine lithium remain heavily dependent on China to convert it into a usable form. Even emerging projects in Europe and North America often rely on Chinese technology or partnerships.

In cobalt, China's influence is even more pronounced. Although most cobalt is mined in the Democratic Republic of Congo, Chinese companies control or finance many of the country's largest cobalt and copper operations. The intermediate cobalt hydroxide from these mines is shipped largely to China, where it is refined into cobalt sulfate or metal. Chinese refiners account for the overwhelming majority of global cobalt chemical production used in batteries. This gives China dual leverage: control over an essential battery material, and influence over the industrial future of regions that depend on cobalt-intensive chemistries.

Nickel provides a different but equally telling example. Indonesia's rise as the world's largest nickel producer was facilitated by Chinese investment and technology. When Indonesia banned raw ore exports to encourage domestic value addition, China's metallurgical companies filled the gap by building smelters, high-pressure acid leaching (HPAL) plants, and refining facilities across the archipelago. These operations produce nickel pig iron, ferronickel, and increasingly nickel sulfate suitable for battery cathodes. Although Indonesian nickel is technically outside China, the technology, capital, and operating expertise remain overwhelmingly Chinese. This has created an extended ecosystem where refining capacity is geographically distributed but strategically aligned with China's industrial base.

Graphite represents one of China's deepest command points in the value chain. The country produces the majority of the world's natural graphite and virtually all battery-grade purified graphite. Purification typically requires processes involving hydrofluoric acid—techniques that China mastered long ago but remain challenging elsewhere due to environmental and regulatory constraints. Even where graphite mines emerge in Africa or North America, the purification stage often leads back to China. Without purified graphite, lithium-ion anodes cannot be produced, making this a critical vulnerability for global battery makers.

China's strength extends further into minor but strategic materials such as gallium and germanium. These elements, essential for semiconductors, optoelectronics, fiber-optic communication, and infrared optics, are mostly recovered from byproduct streams in aluminum and zinc refining. China produces most of the world's gallium and a large share of germanium. The global significance of this dominance became clear when China imposed export licensing requirements on both materials, causing disruptions for semiconductor and high-tech industries worldwide. These measures demonstrated China's ability to wield its refining position as a strategic tool.

Chemical processing capacity is the ultimate foundation of China's power. Refining facilities require massive amounts of electricity, water, labor, specialized chemistry, and technical expertise. Many Western smelters or refineries shut down over the past decades due to environmental concerns or economic pressures. Meanwhile, China expanded capacity across the board: rare earth separation plants, cathode-material factories, anode-material refineries, manganese sulfate facilities, nickel and cobalt refineries, and numerous other chemical processing centers. Once these facilities exist at scale, they create a self-reinforcing loop: downstream manufacturers prefer locating near refined materials, which in turn encourages further growth in refining.

This dominance has profound consequences for global supply chains. It gives China leverage not only during crises but also in day-to-day industrial negotiations. Countries dependent on Chinese materials must consider the political risks of antagonizing China or forming competing alliances. Companies planning battery factories in Europe or North America may find themselves dependent on materials that can only be sourced or processed in China. Defense industries, concerned with rare earth magnet supply, face constraints because Chinese-made magnets cannot be used in sensitive military equipment. Even consumer electronics companies depend heavily on Chinese supply chains for specialty materials, components, and finished products.

China's dominance also affects price formation. Because it can expand or contract production, it can influence global prices for rare earths, cobalt chemicals, graphite anodes, and several other materials. This ability to modulate supply gives China strategic flexibility and allows it to stabilize domestic industries while potentially squeezing foreign competitors. The country's position as a processor and manufacturer also enables it to absorb price shocks more easily than regions that merely import refined materials.

Efforts by other nations to mitigate dependence on China have intensified. The United States, European Union, Japan, South Korea, Australia, and Canada are developing critical minerals strategies centered around reshoring or "friend-shoring" refining capacity. Policies include direct subsidies, tax incentives, streamlined permitting processes, public-private partnerships, and expanded geological surveys. The U.S. has invoked the Defense Production Act to support domestic production of rare earth magnets and battery materials. The EU's Critical Raw Materials Act sets explicit targets for domestic mining, processing, and recycling. Japan continues to invest in overseas projects and stockpiling, while South Korea focuses on securing materials for its battery giants. Yet despite these efforts, scaling refining capacity takes time—often a decade or more—and requires overcoming technical, environmental, and economic obstacles.

China is not static in this landscape. It continues to expand its capabilities, investing in overseas mines, improving domestic refining technologies, and developing sophisticated recycling infrastructure. Chinese companies are acquiring stakes in lithium, nickel, cobalt, and rare earth projects worldwide. They are also moving into regions like Africa, Latin America, Southeast Asia, and even parts of Europe with financing packages, infrastructure development, and long-term offtake agreements. This outward expansion ensures a steady flow of raw materials into China's processing centers while limiting the ability of other countries to develop independent value chains.

China's dominance creates not only vulnerabilities, but also incentives for cooperation. Some countries rely on Chinese partners to develop local industries. Others seek to negotiate more favorable terms for access to processed materials. In some cases, China's control over refining may act as a stabilizing force, preventing extreme price volatility by adjusting production as needed. Yet the risks remain clear: concentration introduces fragility into global systems. A disruption due to internal policy changes, energy shortages, environmental crackdowns, or geopolitical tensions could cascade across multiple industries.

China's position rests less on what it extracts than on how thoroughly it processes, coordinates, and plans across the value chain. This form of dominance is industrial rather than geological, systemic rather than episodic. Any effort to rebalance global supply must therefore contend with capabilities that took decades to assemble and cannot be replicated quickly or selectively.

CHAPTER 14 – ALTERNATIVE PRODUCERS AND EMERGING FRONTIERS

As global industries confront the risks of over-reliance on a single country for critical materials, attention has shifted toward the regions and nations capable of offering real alternatives to China's dominance. The goal is not to replace China entirely—an unrealistic notion given the depth of its industrial ecosystem—but to build a wider, more resilient network of producers across mining, refining, and intermediate manufacturing. This chapter explores the emerging players, the strengths and limitations of each, and the degree to which they could reshape global supply chains in the coming decades.

The first set of alternative producers comprises established resource-rich nations that already play central roles in mining and are now expanding into refining. Australia is the clearest example. It is the world's largest producer of lithium ore and a significant source of rare earths, nickel, and other minerals. For years, much of Australia's raw output was shipped to China for processing. That dynamic is beginning to change. New lithium refineries have been built or planned in Western Australia to produce battery-grade chemicals domestically. Rare earth mining companies are developing separation plants, while partnerships with the United States, Japan, and the European Union aim to create diversified rare earth value chains. Australia's strength lies in its stable political environment, strong regulatory oversight, high environmental standards, and established mining expertise. Its main challenge is cost: energy prices, labor costs, and permitting timelines tend to be higher than in China or Southeast Asia, slowing expansion of downstream processing.

Canada is emerging as another major frontier. With significant deposits of nickel, cobalt, copper, and rare earths—and abundant hydropower for low-carbon refining—Canada is positioning itself as a supplier for the United States and Europe. Nickel and cobalt mines in Ontario and Quebec, along with rare earth projects in the Northwest Territories, have received new investment as Western firms seek cleaner, geopolitically aligned sources of critical materials. Canada is also attracting automotive and battery companies building cathode and cell factories, reinforcing its potential as a regional supply hub. Yet Canada faces its own barriers: high infrastructure costs in remote regions, lengthy permitting processes, and the need for stronger coordination between federal and provincial governments.

Latin America represents a different kind of opportunity. Countries such as Chile, Argentina, and Brazil hold enormous mineral reserves, particularly for lithium, copper, nickel, and rare earths. Chile and Argentina sit atop the lithium-rich salars of the Andes, while Brazil possesses huge niobium resources and growing nickel output. These countries benefit from vast geological endowments and long mining experience. However, they also face political cycles, regulatory uncertainty, social opposition to mining expansion, and debates over resource nationalism. Chile's evolving lithium policy, Argentina's economic instability, and Brazil's environmental governance debates all illustrate the tensions between capturing more value domestically and maintaining investor confidence. If these nations can establish stable regulatory environments and attract responsible refining partners, they could become pillars of a diversified global supply chain.

Africa is another major emerging frontier. The continent is home to rich deposits of cobalt, copper, manganese, graphite, rare earths, and other essential minerals. The Democratic Republic of Congo holds the world's largest cobalt reserves, Botswana and South Africa lead in manganese, Tanzania and Mozambique produce graphite, and several countries—including Malawi, Burundi, Angola, and South Africa—hold promising rare earth deposits. Africa's potential is enormous, but so are the challenges. Infrastructure deficits, political instability, governance issues, and environmental and social concerns can deter long-term investment. Still, there is momentum: new partnerships with the European Union and United States aim to develop processing capacity within Africa rather than exporting raw materials. Zambia and the DRC have announced plans to develop regional battery supply chains. If these initiatives succeed, Africa could shift from being merely a raw material supplier to a player in processing and manufacturing.

Southeast Asia is rapidly becoming a processing hub in its own right, especially Indonesia. The country has leveraged its nickel laterite resources by banning raw ore exports and encouraging the construction of domestic refining plants. Chinese companies have heavily invested in smelters, HPAL facilities, and battery precursor plants. Although Indonesia's progress is tied to Chinese capital and technology, it nonetheless represents a geographic diversification of nickel refining capacity. The Philippines remains an important supplier of nickel ore, though environmental regulations have limited expansion. Vietnam, with emerging rare earth deposits and refining ambitions, is partnering with Japanese and Korean companies to upgrade its industrial base.

India presents one of the most intriguing long-term opportunities. With rare earth deposits, titanium, bauxite, and growing interest in battery materials, India has the geological foundation for a more prominent role. The country has articulated ambitions to develop its own rare earth supply chain and expand into advanced materials manufacturing. Yet progress has been slow. India faces challenges similar to those of other democracies: permitting constraints, environmental concerns, bureaucratic complexity, and the need for capital-intensive chemical plants. However, with rising geopolitical alignment among India, the United States, Japan, and Europe, investment flows may accelerate. India could eventually become a significant refining and manufacturing hub—if it addresses regulatory bottlenecks and builds the necessary expertise.

Europe faces a different challenge: it is not rich in mineral resources, yet it is determined to secure its own supply chains for energy transition technologies. Though it lacks the geological abundance of Australia or Africa, Europe does possess deposits of lithium, cobalt, nickel, and rare earths. Projects in Sweden, Finland, Portugal, Spain, and Germany aim to revive mining, but these face community opposition, environmental scrutiny, and long approval timelines. Europe's real strength lies in its industrial base, technological expertise, and strong regulatory frameworks. It is investing heavily in refining capacity, battery materials, and recycling infrastructure. European Union policies now target domestic processing and reduced dependency on single-country imports. If the EU succeeds in building a recycling-centered supply chain—leveraging high recovery rates for cobalt, nickel, and copper—it may partially offset its geological constraints.

Japan and South Korea remain technologically sophisticated actors with advanced refining capabilities in specific materials, although their processing capacity cannot match China in scale. Japan is highly skilled in rare earth magnet manufacturing and has pioneered recycling technologies. South Korea excels in battery materials, especially cathodes, and is expanding partnerships with global mining firms. Both nations invest strategically in overseas mines to secure supply, often in collaboration with allies. Their strength lies not in raw production but in knowledge, reliability, and high-quality processing.

New frontiers are also emerging in unexpected places. The Middle East, traditionally associated with oil and gas, is increasingly investing in critical minerals. Saudi Arabia, the United Arab Emirates, and Oman have announced ambitious plans to develop mining sectors, refine strategic materials, and attract global industrial partners. These countries offer energy advantages—low-cost solar power, future hydrogen production—and strong financial resources. While their geological potential is still being assessed, their capital and political drive could make them important refining centers for minerals sourced elsewhere.

Another frontier lies in technological innovation rather than geography. Companies worldwide are developing advanced refining methods such as direct lithium extraction, membrane-based rare earth separation, solvent extraction alternatives, and cleaner graphite purification. Successful commercialization of these technologies could enable new regions to enter the refining stage without replicating the environmental impacts associated with traditional methods. Recycling forms another technological frontier. Over time, end-of-life batteries, magnets, electronics, and catalysts will become valuable urban mines. Nations that establish strong recycling industries could become major suppliers of secondary materials, reducing the need for new extraction and diversifying supply chains.

Despite these opportunities, replacing or significantly reducing dependence on China remains difficult. No single country can replicate China's scale, cost structure, or integrated ecosystem across mining, processing, component manufacturing, and end-use industries. Even groups of countries must coordinate extensively to build parallel supply chains. Diversification will not eliminate interdependence; rather, it will distribute it more evenly across a broader array of producers.

Still, the momentum is unmistakable. Governments and companies are now pursuing long-term strategies to reshape supply chains, driven by geopolitical risk, energy transition goals, and the desire to capture more economic value from critical materials. The process will be uneven, shaped by domestic politics, global market cycles, and technological breakthroughs. Some nations will emerge as major new producers; others will find their ambitions constrained by environmental concerns or limited capacity. But over the coming decade, diversification efforts will meaningfully reduce the concentration risk that has defined the last twenty years.

Emerging producers enter markets shaped by powerful economic forces. Supply constraints, demand elasticity, and capital intensity determine why some materials remain structurally scarce while others oscillate between surplus and shortage. These dynamics define the boundaries within which material strategy can operate.

CHAPTER 15 – THE ECONOMICS OF SUPPLY AND DEMAND

The economics of critical materials differ profoundly from those of traditional commodities. Unlike oil, coal, or iron ore—whose markets are shaped by high volumes, broad substitutability, and mature global systems—critical raw materials operate in markets defined by geological scarcity, technological specificity, volatile demand, long development timelines, and highly concentrated supply chains. Their economics are not driven by simple price cycles but by structural constraints, strategic behavior, and geopolitical dynamics. Understanding these forces is essential for anticipating shortages, managing industrial risk, and designing effective policy.

At the heart of critical material economics lies asymmetry: demand is global, fast-growing, and technology-driven, while supply is slow, capital-intensive, and geographically concentrated. This asymmetry generates persistent tension. New demand emerges rapidly as industries scale electric vehicles, wind turbines, solar panels, data centers, semiconductors, and advanced defense systems. Yet new mines, refineries, and processing facilities take years—often more than a decade—to reach commercial production. As a result, demand curves often rise much faster than supply capacity can adjust, creating structural tightness in several markets.

This dynamic has become especially acute for lithium, nickel, cobalt, and rare earths. The energy transition is reshaping global demand patterns for these materials. Electric vehicles alone are projected to account for a substantial portion of future lithium, nickel, and cobalt consumption. Wind turbines require rare earth magnets, and grid expansion requires copper and aluminum. Because these technologies are scaling at unprecedented speed, demand forecasts have grown more ambitious in each successive year. For many critical materials, demand is expected to double, triple, or even quadruple before 2040. These trajectories create enormous economic pressure on suppliers.

However, supply cannot simply "scale on demand." Mining projects face exploratory uncertainty, geological constraints, and permitting delays. Refining plants require specialized expertise, stable energy, and substantial capital. Even small disruptions—a regulatory delay, a community dispute, a change in environmental standards—can push projects years behind schedule. Once operational, mines and refineries must navigate fluctuating prices, shifting regulatory regimes, and geopolitical risks. This combination of long lead times, high capital requirements, and unpredictable markets creates chronic underinvestment, especially during price downturns. When demand rebounds, supply lags—producing familiar boom-and-bust patterns.

Yet the boom-and-bust cycles of critical materials differ significantly from those of traditional commodities. In many cases, demand is not cyclical but structural. The push toward electrification, digitalization, and clean energy is not expected to reverse. Even when prices spike, demand does not fall as sharply as it might for discretionary commodities because manufacturers cannot easily switch to alternative materials. An electric vehicle manufacturer cannot simply decide to produce fewer cars when lithium prices rise; instead, it faces higher battery costs and tighter margins. This inelasticity means price volatility disproportionately affects downstream sectors while offering limited signals for upstream investment.

Inelasticity also exists on the supply side. Mines cannot ramp up output quickly in response to high prices. They cannot easily reduce production during price drops, either, because fixed costs remain high. This creates conditions where prices can become detached from short-term fundamentals, driven instead by scarcity premiums or speculative movements. Lithium's dramatic price spikes in 2021–2022 showed how demand surges, limited refining capacity, and investor speculation can combine to produce extreme volatility. When new supply eventually enters the market, prices can collapse, discouraging further investment and setting the stage for future shortages—a classic resource paradox.

Another defining feature of critical material economics is concentration. In many markets, one or a few countries dominate production and refining. Concentration alters price formation, market behavior, and investment flows. For rare earths, China's control over separation and refining gives it outsized influence over global prices, even when mining diversifies. For cobalt, the dominance of the Democratic Republic of Congo creates supply risks that cannot be smoothed out by diversification alone. Indonesian nickel policy—particularly its export bans and incentives for domestic processing—has restructured global market dynamics in a matter of years. Concentration can also suppress price transparency, as many critical materials are traded through opaque long-term contracts rather than open exchanges.

Geopolitical dynamics amplify these economic complexities. Export controls, sanctions, strategic stockpiling, and industrial policy all affect supply and demand. China's export licensing requirements for gallium and germanium demonstrated how political decisions can disrupt global supply chains for niche materials essential to semiconductors and optics. Indonesia's nickel export ban illustrated how nations can use resource leverage to attract investment and promote domestic value addition. Western governments are increasingly intervening in critical material markets through subsidies, tax credits, and strategic reserves. These interventions shape market behavior, often pushing companies to prioritize security of supply over short-term price considerations.

Meanwhile, industrial policy has become a powerful economic driver. The United States' Inflation Reduction Act, the European Union's Critical Raw Materials Act, Japan's strategic stockpiling programs, and South Korea's battery-material investments all aim to reshape supply chains through targeted incentives. These policies create new demand signals and alter investment flows. They also introduce competition among allies, each seeking to build domestic processing capacity and secure stable access to raw materials. This competition can distort markets but also accelerate diversification.

A critical factor in understanding supply-demand dynamics is the role of technology. Technological innovation affects both sides of the equation. On the demand side, new battery chemistries influence material composition. Cathode evolution—from NMC to NCA to high-nickel or cobalt-free variants—changes the relative importance of nickel, cobalt, and manganese. Lithium iron phosphate (LFP) reduces cobalt and nickel demand but increases graphite demand. Sodium-ion batteries could reduce lithium intensity in certain applications, though their energy density limitations constrain their use. Rare earth-free motors, alternative magnets, and improved recycling technologies also reshape demand patterns. However, technological shifts rarely eliminate demand entirely; rather, they redistribute it across the periodic table.

On the supply side, new extraction and processing technologies may unlock previously uneconomic resources. Direct lithium extraction, advanced rare earth separation methods, novel leaching techniques, and more efficient refining chemistries hold the potential to lower costs or expand supply. But commercialization is slow, capital-intensive, and subject to technical hurdles. These innovations may eventually ease supply constraints, but they cannot fully solve the structural lag between demand and supply.

Recycling is often presented as a solution to long-term imbalance, but its role is nuanced. Today's recycling rates for critical materials such as lithium, rare earths, and graphite remain low due to technological, economic, and logistical barriers. Moreover, the installed stock of batteries, turbines, electronics, and magnets is still relatively young. Significant volumes of end-of-life material will not return to the supply cycle until the 2030s. When it does, recycling could fundamentally reshape supply-demand economics—but not soon enough to mitigate near-term shortages. For now, secondary supply helps more for metals like copper, aluminum, and precious metals, where recycling infrastructures are mature.

A further layer of complexity is introduced by material substitutability—or the lack thereof. Some critical materials have few or no viable substitutes. Rare earth magnets are essential for high-performance electric motors and wind turbines; gallium and germanium are critical for advanced semiconductors; graphite remains unmatched in lithium-ion anodes. When materials cannot be substituted, prices can spike without meaningfully reducing demand. This exacerbates volatility and challenges the efficiency of traditional market mechanisms.

The economics of critical materials also reflect the importance of long-term contracts. Many producers and buyers prefer structured agreements over spot transactions to manage volatility and secure supply. These contracts influence project financing, shape price expectations, and determine which mines or refineries receive investment. The prevalence of long-term contracts also means that spot-market prices often represent only a fraction of the real economic landscape.

Ultimately, critical material markets operate in a hybrid space—part geology, part chemistry, part geopolitics, and part industrial policy. Supply-demand relationships cannot be understood solely through classical economic models because they involve factors that defy short-term elasticity and traditional competition. These markets respond not just to prices but also to technological trajectories, government strategies, investor sentiment, and strategic rivalry.

CHAPTER 16 – RISKS, VULNERABILITIES, AND DISRUPTIONS IN GLOBAL CHAINS

Critical material supply chains are among the most complex, fragile, and strategically exposed systems in the global economy. They stretch across continents, involve intricate chemical processing, depend on specialized infrastructure, and rely heavily on a small number of producers and refiners. This combination creates structural vulnerabilities that can propagate disruptions quickly and unpredictably. As the energy transition accelerates and demand for critical materials grows, understanding these risks becomes essential not only for governments and corporations, but for the stability of entire technological systems.

The first and most fundamental vulnerability is concentration. For many critical materials, extraction and refining are dominated by one or two countries. Rare earths are overwhelmingly refined in China. Cobalt comes primarily from the Democratic Republic of Congo. Nickel refining is concentrated in Indonesia and China. Lithium refining is dominated by China. Platinum group metals are largely sourced from South Africa and Russia. Graphite purification is almost entirely located in China. These concentrations mean that a disruption in a single region—whether due to political instability, export controls, natural disasters, or environmental enforcement—can cascade through global industries.

The fragility of this concentration becomes clear when examining past disruptions. When China restricted rare earth exports to Japan in 2010, prices spiked dramatically and manufacturers scrambled to redesign supply chains. More recently, China's decision to impose licensing requirements on gallium and germanium sent ripples through semiconductor and optics supply chains, revealing how smaller-volume materials can become chokepoints for critical industries. Energy shortages in China in 2021 shut down magnesium and silicon metal plants, crippling downstream aluminum producers in Europe. These events show that supply chain fragility is not theoretical; it is structural and recurring.

Political risk is deeply intertwined with concentration. Countries dominating critical material supply often face governance challenges or geopolitical tensions. In the Democratic Republic of Congo, political instability, conflict mineral concerns, and social issues can disrupt cobalt mining. Indonesia's policy shifts—including export bans and domestic processing mandates—have repeatedly reshaped nickel markets. South Africa's recurring electricity shortages undermine platinum and palladium production. Russia's war in Ukraine raised concerns about palladium, titanium, and other materials linked to Russian supply. These political dynamics influence not only current output but also long-term investment, creating uncertainty that deters diversification.

Environmental and regulatory risks form another major vulnerability. Critical material extraction and refining are often environmentally intensive, involving high water consumption, chemical waste, energy use, or radioactive byproducts. As environmental regulations tighten—especially in Europe, North America, and parts of Asia—projects face increasing scrutiny and longer permitting processes. These regulatory hurdles slow new supply even as demand rises. The paradox is clear: societies demand cleaner technologies but resist the mining and refining necessary to build them. Permitting for new mines in Europe or the United States can take a decade or more, leaving the world dependent on producers with weaker environmental safeguards. When environmental authorities in any major producing country tighten controls—as seen in China's rare earth sector—the global impact is immediate.

Energy itself is an overlooked but critical source of vulnerability. Many refining processes require enormous amounts of electricity or heat. Aluminum smelting, nickel refining, magnesium production, and rare earth separation all depend heavily on stable, affordable energy supplies. When energy prices spike, refineries shut down. When electricity grids become unstable, production halts. The European energy crisis of 2021–2022 caused widespread closures of metal smelters, contributing to global shortages of magnesium, zinc, and silicon metal. This demonstrated how energy transitions, energy policies, and geopolitical energy shocks directly affect material supply chains. In a decarbonizing world, regions with access to abundant renewable energy may have a competitive advantage in refining—while regions dependent on fossil-intensive grids may lose ground.

Logistical risk adds another layer of fragility. Critical materials often travel long distances between mine, refinery, chemical plant, component factory, and final manufacturer. These routes depend on shipping lanes, rail networks, and port capacity. The COVID-19 pandemic exposed the vulnerability of global logistics, with port closures, container shortages, and shipping delays affecting everything from battery materials to semiconductor equipment. Even temporary disruptions—such as the blockage of the Suez Canal—can delay shipments, increase costs, and force manufacturers to slow production. Given the geographic distribution of critical material supply chains, logistical disruptions can amplify shortages caused by mining or refining constraints.

Price volatility itself is a vulnerability. Critical materials markets are often small, opaque, and dominated by long-term contracts rather than transparent exchanges. Prices can swing dramatically in response to small changes in supply or demand. Lithium, nickel, cobalt, and rare earths have each experienced periods of extreme volatility. For manufacturers, this volatility makes long-term planning difficult. For investors, it increases uncertainty, discouraging investment in new capacity. For governments, price spikes can threaten energy transition goals by making electric vehicles, batteries, and renewables more expensive. Volatility discourages the development of stable supply chains and can undermine industrial strategy.

Supply-demand mismatches create structural risk. Many critical materials face tight supply that cannot keep pace with rapid demand growth. Lithium demand is soaring due to electric vehicles. Rare earth magnet demand is rising due to wind turbines, robotics, and electric motors. Copper demand is projected to surge as grid infrastructure expands. Yet new mining and refining capacity for these materials is slow to develop. Even optimistic projections suggest that supply will struggle to meet demand for several years. This structural gap creates vulnerability across industries, from automotive to aerospace to renewable energy. If supply falls short, nations may be forced to prioritize certain technologies over others, slowing climate progress or reducing industrial competitiveness.

Another vulnerability lies in the byproduct dependency of many critical materials. Germanium depends on zinc smelting. Gallium depends on alumina refining. Tellurium depends on copper refining. Dysprosium depends on specific rare earth deposits. These dependencies mean that supply is constrained not by demand for the critical material itself, but by the economics of the primary metal. If zinc production declines due to low prices or energy shocks, germanium supply shrinks regardless of demand. If copper refineries slow due to regulatory constraints, tellurium availability drops. This byproduct dependence makes supply unpredictable and difficult to scale, contributing to recurring shortages.

Corporate concentration introduces additional risk. Many critical material markets are dominated by a small number of companies. Rare earth processing involves a handful of Chinese firms. Lithium refining includes a small group of Chinese and Western companies. Nickel laterite refining is dominated by Chinese-Indonesian joint ventures. Concentrated corporate control can reduce market competition, limit transparency, and increase exposure to company-specific disruptions such as labor strikes, financial instability, or operational failures.

Technological vulnerability is also significant. Advanced technologies require highly specific material forms—battery-grade lithium, high-purity nickel sulfate, chemically consistent cathode precursors, precise crystal structures for semiconductors, and ultra-pure rare earth oxides. Producing these materials requires technical expertise that only a few companies possess. A failure at any stage of processing can render a material unusable. Even slight impurities can degrade battery performance, semiconductor reliability, or magnet strength. The complexity of these requirements increases the number of potential failure points in the supply chain.

Cybersecurity risks are growing as well. Mines, smelters, refineries, and battery gigafactories depend on digital control systems. A cyberattack on a major processing facility could halt production and cause global ripple effects. As supply chains become more automated and interconnected, the risk of digital disruption grows.

Taken together, these vulnerabilities produce a supply chain architecture that is fragile by design. Material flows are shaped as much by logistics and trade routes as by geology or production capacity. In this configuration, access depends on who controls the corridors through which resources move.

CHAPTER 17 – LOGISTICS, TRADE ROUTES, AND ENERGY TRANSITION MATERIALS

Critical materials do not move through the world the way traditional commodities do. Oil travels in massive tankers, coal moves in bulk shipments, grain flows through predictable corridors. But the materials that power electrification, digitalization, and advanced manufacturing travel through far more intricate, fragile, and interdependent routes. Their flows depend on long-distance transport, specialized infrastructure, chokepoint geographies, and geopolitical stability. Logistics is not merely a technical detail—it is a strategic layer of the global material system. Understanding these routes reveals how vulnerabilities in transport networks shape the security of the entire energy transition.

Critical material supply chains often stretch across continents multiple times before they enter a finished product. Lithium mined in Australia may travel to China for refining, to South Korea for cathode production, to Europe for battery cell manufacturing, and to the United States for installation in an electric vehicle. Rare earth concentrates may move from Myanmar to southern China for separation, then to Japan for magnet manufacturing, and finally to Europe for wind turbine assembly. Nickel ore from Indonesia may be smelted locally, refined in China, turned into precursor materials in Korea, and used in battery factories in Germany. This global choreography depends on reliable maritime routes, efficient ports, robust container availability, and stable geopolitical conditions.

Maritime logistics dominate the flow of critical materials. More than 80 percent of global trade by volume moves by sea, and critical minerals are especially dependent on maritime corridors. The Indo-Pacific region plays an outsized role. The South China Sea, the Strait of Malacca, and surrounding waters are major arteries for rare earths, cobalt intermediates, nickel products, lithium chemicals, and semiconductor materials. Most refined battery materials come from China, South Korea, and Japan, all of which rely on the safe passage of ships through contested waters. Any disruption—whether from territorial tensions, military activity, piracy, or natural disasters—could delay shipments and choke off supplies for downstream industries.

Chokepoints such as the Strait of Malacca, the Strait of Hormuz, the Suez Canal, the Panama Canal, and the Bab el-Mandeb Strait play disproportionate roles in critical material trade. The Strait of Malacca, one of the busiest shipping lanes in the world, handles a large share of freight between East Asia and Europe. The Suez Canal carries significant volumes of industrial materials and finished components to Europe and North America. The Panama Canal, important for trade between East Asia and the eastern United States, has faced capacity restrictions due to drought—highlighting how climate change introduces new logistical risks. These chokepoints compress global trade into narrow physical corridors where disruption can ripple across global value chains.

Port infrastructure forms another essential component of logistics. Ports in China—Ningbo, Shanghai, Tianjin, and Shenzhen—are central to the flow of refined materials and intermediate products. Busan in South Korea and Yokohama in Japan serve as critical transshipment hubs for battery and semiconductor supply chains. Australian ports such as Port Hedland and Fremantle handle significant volumes of lithium and nickel. Rotterdam, Antwerp, and Hamburg anchor Europe's import hubs for battery materials and advanced manufacturing inputs. Weaknesses in port capacity, labor strikes, customs delays, or regulatory upgrades can slow the flow of goods and create backlogs that affect industries thousands of miles away.

The COVID-19 pandemic demonstrated how sensitive critical material supply chains are to logistical shocks. Port shutdowns in China, factory closures in Southeast Asia, and container shortages disrupted the flow of key materials. Electronics manufacturers experienced delays in rare earth components and semiconductor inputs. Battery manufacturers faced interruptions in shipments of cathode precursors and graphite anodes. The pandemic exposed how "just-in-time" logistics, optimized for efficiency rather than resilience, leave supply chains vulnerable to systemic disruptions. Even a temporary mismatch in container availability or shipping schedules can slow global production.

Air freight plays a less visible but critical role for certain high-value materials. Semiconductors, specialized chemicals, micro-components, and certain metal powders often travel by air due to their sensitivity or value density. Disruptions to air cargo routes—whether from geopolitical tensions, sanctions, weather, or pandemic restrictions—can delay production for industries that rely on rapid replenishment cycles. While bulk materials like copper, nickel, and lithium salts almost always move by sea, the highest-value segments of the supply chain depend on air routes that can shift quickly in response to crises.

Rail systems account for a smaller portion of critical material trade but have strategic importance. The rail corridors connecting China to Europe—often referred to as the Belt and Road rail routes—move significant volumes of components and intermediate goods. These routes gained attention during maritime disruptions, offering an alternative for time-sensitive shipments. However, geopolitical tensions, sanctions, and instability in transit regions (such as Russia or Central Asia) complicate their long-term reliability. Rail may complement maritime shipping, but it cannot replace it on scale.

Energy transition materials impose new demands on global logistics. Copper, lithium, cobalt, nickel, rare earths, and graphite must move in volumes far greater than historical levels. Many supply chains lack sufficient port capacity, storage facilities, rail links, and handling equipment to accommodate the coming surge. If the world succeeds in scaling electric vehicles, renewable energy, and battery storage, the movement of critical materials could expand by orders of magnitude. Without upgraded infrastructure, bottlenecks may shift from mines and refineries to ports and shipping lanes.

Climate change adds a destabilizing force. Drought conditions have reduced the capacity of the Panama Canal, slowing global trade. Rising sea levels threaten port infrastructure. More frequent storms disrupt shipping routes, damage port facilities, and delay deliveries. Extreme heat affects rail operations and storage conditions for sensitive chemicals. As climate impacts intensify, logistics networks must adapt—or risk becoming another source of disruption.

Geopolitical tensions increasingly intersect with trade routes. Great-power rivalry between the United States and China reshapes supply chain planning. Countries are beginning to rethink reliance on East Asian shipping routes for essential materials. Initiatives such as the Indo-Pacific Economic Framework, the India–Middle East–Europe

Corridor, and the EU's Global Gateway aim to create alternative trade routes and reduce dependence on contested maritime zones. At the same time, China continues to expand its influence over global ports through overseas investments, especially via its Belt and Road Initiative. These investments give China strategic visibility and leverage over logistics networks that other nations depend upon.

Sanctions and export controls present another layer of risk. Restrictions on Russian palladium, titanium, and nickel altered flows and forced producers to reconfigure supply chains. Export controls on semiconductor materials affect shipments of gallium, germanium, and advanced chemical precursors. New regulatory regimes in the United States, Europe, and Japan scrutinize the movement of sensitive technologies and materials. Companies now operate in a landscape where geopolitical decisions can redraw logistics routes abruptly.

Insurance and financial services also shape material flows. Shipping insurance rates rise during conflicts, piracy events, or geopolitical instability. Higher rates increase the cost of moving critical materials and may deter shipments through risky corridors. In extreme cases, insurers may refuse coverage for certain routes, forcing ships to take longer paths that increase transit times and emissions.

Security considerations amplify vulnerabilities. Piracy remains a threat in regions such as the Gulf of Guinea and the Strait of Malacca. Sabotage, terrorism, and cyberattacks on port infrastructure pose rising risks. As more ports digitize operations, cyber vulnerabilities can shut down terminals or compromise cargo-tracking systems.

Efforts to build resilience in logistics are accelerating. Diversified shipping routes, increased inventory buffers, multi-source procurement strategies, and the development of regional supply hubs are becoming more common. Battery manufacturers are increasingly siting production in proximity to refining facilities and automotive clusters to reduce transport dependencies. Some countries are investing in strategic reserves of critical materials not just to buffer supply shocks but also to smooth logistical disruptions. New trade agreements seek to harmonize customs procedures and reduce friction at borders.

Recycling and circularity could eventually reshape logistics by reducing the need to move raw materials across long distances. Urban mining facilities near major industrial centers may extract lithium, cobalt, nickel, and rare earths from end-of-life batteries, electronics, and motors. This would shorten supply chains, lower shipping volumes, and reduce the environmental impact of material transport. However, recycling is not yet mature enough to offset primary supply needs.

In the decades ahead, logistics will become an essential dimension of geopolitical strategy. Nations that secure reliable trade routes, resilient infrastructure, and diversified transport networks will gain significant advantages. Those that rely on vulnerable chokepoints or politically unstable regions will face recurring disruptions. The energy transition requires not just geologically abundant materials but the ability to move them reliably—across oceans, borders, and industrial ecosystems.

Material security now shapes how states compete, cooperate, and position themselves within the international system. Control over supply chains feeds directly into alliance structures, strategic autonomy, and bargaining power. In this environment, critical materials are no longer an industrial concern but a central variable of geopolitics.

PART III – STRATEGIC SECTORS AND INDUSTRIAL USES
CHAPTER 18 – ELECTRIC VEHICLES: BATTERIES, MOTORS, AND MATERIALS

The electric vehicle revolution is not simply a story about transportation; it is a story about materials. Every EV embodies a dense concentration of critical minerals, advanced chemistries, rare earth magnets, semiconductor components, and high-performance alloys. Unlike internal combustion vehicles, which rely heavily on mechanical engineering and fossil fuels, EVs depend on a new industrial ecosystem built around minerals extracted, refined, processed, and assembled across complex global supply chains. Understanding these material foundations is essential to understanding the challenges and opportunities of electrification.

Batteries are the beating heart of the electric vehicle. Their chemistry determines range, cost, safety, charging speed, and even the economic competitiveness of entire automotive industries. Lithium-ion technology dominates today's EV market, though within this category a variety of chemistries shape demand for different materials. Nickel-manganese-cobalt (NMC) batteries, commonly used in long-range EVs, rely on high-purity nickel to increase energy density, cobalt to stabilize the structure, and manganese to balance cost and performance. Lithium-nickel-cobalt-aluminum (NCA) chemistries follow a similar pattern, emphasizing nickel content for high-performance applications. Lithium iron phosphate (LFP) batteries, which avoid nickel and cobalt, are growing rapidly due to cost advantages and thermal stability, especially in mass-market vehicles. Despite these differences, all lithium-ion batteries require lithium, graphite anodes, copper foils, aluminum casings, and a suite of electrolyte chemicals and separators.

Lithium demand has surged alongside EV adoption. Hard-rock spodumene from Australia and brine-based lithium from the Andes are refined primarily in China, where battery-grade carbonate and hydroxide production is concentrated. These compounds must meet extremely strict purity requirements. Even trace impurities can degrade battery performance, shorten lifespan, or create safety risks. This sensitivity makes lithium refining a critical chokepoint in EV supply chains. As automakers scale production, they face not only the challenge of securing lithium but of securing refined lithium of the right chemical quality.

Nickel demand is equally central, especially for high-energy-density batteries. Class 1 nickel, required for battery cathodes, is far more difficult and expensive to produce than lower-grade nickel used in steelmaking. Indonesia's nickel laterite deposits have become the primary upstream source, but converting laterite into battery-suitable nickel sulfate requires energy-intensive high-pressure acid leaching. China has invested heavily in Indonesian processing plants, creating a vertically integrated nickel-to-battery ecosystem that is difficult for competitors to replicate. This interdependence means that global EV production is tied to the environmental, political, and energy conditions of a handful of Southeast Asian and Chinese industrial regions.

Cobalt, while being phased down in many cathode chemistries, remains indispensable in stabilizing high-energy cathodes. The Democratic Republic of Congo produces most of the world's cobalt, and Chinese refiners convert most of that into battery chemicals. This dual concentration creates strategic vulnerabilities. Supply disruptions, political instability, or regulatory shifts can influence global EV manufacturing more directly than consumers realize. Even as automakers adopt lower-cobalt or cobalt-free chemistries, the legacy of cobalt dependence lingers in supply agreements, refinery contracts, and ongoing reliance on mixed cathode portfolios.

Graphite is another foundational material, though its profile is less publicly recognized. Natural and synthetic graphite form the anode of nearly every lithium-ion battery. Purified graphite must reach extremely high carbon purity and precise particle characteristics, a process overwhelmingly conducted in China. Despite emerging mines in Africa and synthetic graphite production in other countries, China's dominance in purification represents one of the EV sector's most significant bottlenecks. Without purified graphite, batteries cannot be produced at scale.

Beyond batteries, the electric motor is another arena where materials shape performance and industrial strategy. Most high-efficiency EV motors rely on permanent magnets composed of neodymium, praseodymium, and dysprosium. These rare earth magnets generate strong, efficient fields that allow motors to deliver high torque at low weight. Their unique electron structures produce magnetic properties unmatched by other materials. While induction motors and wound-field motors exist as rare-earth-free alternatives, permanent magnet motors remain the preferred technology for efficiency-driven vehicle architectures. Rare earth supply therefore exerts major influence over EV production, with China supplying most of the world's separated heavy rare earths and dominating magnet manufacturing.

Copper is another silent giant of the EV revolution. EVs require far more copper than internal combustion vehicles—used in wiring harnesses, power electronics, inverters, charging systems, and motors. As electrification expands across transportation, grid infrastructure, renewable energy, and charging networks, global copper demand is set to rise dramatically. Yet new copper mines face long permitting timelines, declining ore grades, and environmental constraints. This mismatch between rising demand and limited new supply highlights how EVs influence minerals far beyond their own components.

Power electronics form another critical subsystem. Silicon carbide (SiC) and gallium nitride (GaN) semiconductors enable fast charging, higher efficiency, and improved thermal management. As EV drivetrains become more advanced, demand for these compound semiconductors will grow. These materials require specialty inputs such as high-purity silicon carbide powder, gallium, and rare earth dopants. Their supply chains, heavily concentrated in a few countries, represent another potential risk as EV adoption accelerates.

Charging infrastructure introduces further material dependencies. High-power chargers require copper, aluminum, and specialty steels. Data connectivity, sensors, and grid-management technologies rely on rare metals, high-purity silicon, and advanced semiconductor devices. As fast charging expands globally, demand for these supporting materials will increase, creating competition with other industries such as data centers, telecommunications, and renewable energy systems.

The economics of EV materials are not purely about securing supply; they are also about developing stable, predictable, long-term cost structures. Automakers entering the EV market confront a commodity landscape far more volatile than the petroleum market. Lithium prices have spiked and collapsed repeatedly. Nickel prices have seen extreme volatility due to market manipulation, policy shifts, and supply disruptions. Rare earth prices can be influenced by export restrictions, energy shortages, or regulatory crackdowns in China. These fluctuations complicate cost modeling and require automakers to adapt procurement strategies, including long-term contracts, joint ventures with miners and refiners, and— increasingly—direct investments in upstream assets.

Recycling will eventually play a significant role in EV material supply, but its impact will not peak until large volumes of EV batteries reach end of life. Current recycling capacity remains limited, though several companies are developing advanced processes to recover lithium, nickel, cobalt, graphite, and other components. Closed-loop recycling could dramatically reduce material demand in the long term and mitigate geopolitical risks. However, recycling economics are complex, and today's battery designs are not optimized for disassembly or material recovery. Standardization and regulatory incentives will be needed to accelerate the recycling ecosystem.

The industrial geography of EV production is shifting in response to material constraints. Countries attempting to build domestic EV industries must also build or secure reliable supply chains for battery materials, rare earths, copper, and semiconductors. The United States, European Union, Japan, South Korea, and India are investing in domestic battery plants, critical mineral maps, and processing facilities. Yet despite these efforts, it is unlikely that any single region will become fully independent. The EV sector is inherently global, with supply chains spanning continents and involving multiple layers of specialization. The challenge is not to sever global interdependence but to rebalance it, reducing vulnerabilities while maintaining efficiency.

Policy is now one of the most powerful forces shaping EV material demand. Incentives for EV adoption, subsidies for battery factories, restrictions on high-carbon materials, and industrial strategies targeting the energy transition all influence which materials scale, where supply chains form, and how companies plan long-term capacity. Regulatory frameworks such as the Inflation Reduction Act or the European Critical Raw Materials Act aim to localize production and reduce reliance on unstable or unfriendly suppliers. These policies could redefine the competitive landscape for EV materials over the next decade.

The strategic significance of EV materials extends beyond industrial supply. They influence trade relationships, diplomatic alliances, environmental policies, and national security strategies. Nations that control refining or magnet manufacturing can wield influence over automotive industries worldwide. Conversely, nations that depend heavily on imported materials must build new forms of resilience—redundant supply chains, strategic stockpiles, recycling ecosystems, and coordinated trade agreements.

The electric vehicle is therefore not merely a consumer product; it is a geopolitical system on wheels. Inside every battery pack lies a map of global interdependence, of mineral wealth and scarcity, of technological expertise and industrial strategy. As electrification expands into buses, trucks, two-wheelers, and industrial machinery, the pressures on this material system will intensify. The choices made today—in technology, policy, sourcing, and innovation—will shape not only the automotive industry but the politics of energy and materials for decades.

Critical materials extend far beyond vehicles, forming the backbone of the wider energy transition. Wind power, solar generation, and grid infrastructure all depend on material inputs that are both specialized and constrained. The pace and stability of decarbonization are therefore inseparable from the availability and control of these resources.

CHAPTER 19 – WIND, SOLAR, AND THE ENERGY TRANSITION

The global energy transition is fundamentally a materials transition. Wind turbines, solar panels, transmission lines, inverters, and energy-storage systems all rely on minerals and refined materials that enable high performance, durability, and large-scale deployment. These technologies symbolize a clean future, yet their physical foundations tie them to complex supply chains, concentrated refining hubs, and geopolitical competition. Understanding the material intensity of wind and solar—along with the constraints and risks embedded in their supply chains—is essential for understanding the pace, cost, and strategic challenges of decarbonization.

Wind energy is one of the most mineral-intensive forms of power generation. Modern wind turbines are sophisticated machines consisting of towers, nacelles, blades, gearboxes, generators, and control systems—each built from a combination of steel, copper, rare earth magnets, fiberglass, composites, and specialty alloys. Steel dominates the tower structure, accounting for a large share of total mass. Copper is essential for generators, cables, transformers, and grid interconnections. Aluminum appears in various components, contributing lightness where weight matters. But the most strategically important materials in wind turbines are rare earth elements, which underpin high-performance permanent magnet generators.

Direct-drive wind turbines, especially in offshore environments, rely on neodymium-iron-boron magnets doped with dysprosium or terbium for thermal stability. These magnets provide high torque density and efficiency while eliminating the need for large gearboxes that are prone to wear, noise, and maintenance challenges. Offshore turbines, which operate under harsh conditions, depend on these magnets for reliability and low lifecycle costs. As turbine sizes grow—offshore units now reaching 14–18 megawatts—the quantity of rare earth magnets required increases accordingly. Although some onshore turbines use induction generators or geared systems, the industry's movement toward larger and more efficient turbines makes rare earth demand increasingly unavoidable.

This growing reliance intersects with a supply chain that remains heavily concentrated. China dominates rare earth mining, separation, and magnet manufacturing. Even when rare earth ores originate in Australia, the United States, or Myanmar, separation and magnet production usually occur in China. This creates a strategic vulnerability: as wind demand grows, access to rare earth magnets becomes a limiting factor. Attempts to develop rare-earth-free generator technologies continue, but none have matched the performance, efficiency, or reliability of permanent magnet systems at large scale.

Solar photovoltaics present a different material profile but an equally important set of dependencies. Crystalline silicon technology dominates the global market, and its supply chain revolves around quartz, silicon metal, polysilicon, silver paste, aluminum frames, glass, and various dopants. High-purity polysilicon is produced through energy-intensive chemical processes requiring consistent power and sophisticated industrial equipment. Today, China produces the majority of the world's polysilicon, as well as much of the wafer and cell production capacity. The cost advantages of clustered manufacturing, access to low-cost energy, and large-scale industrial policy investments have consolidated the industry's center of gravity in East Asia.

Silver is one of the most critical materials in current photovoltaic manufacturing. Each solar cell uses small amounts of silver paste to form conductive contacts that extract current. At scale, this makes solar a major consumer of global silver production. Efforts are underway to reduce silver intensity through copper substitutes or novel metallization techniques, but progress is gradual. As solar deployment accelerates, tension between silver supply and photovoltaic demand will intensify unless alternatives reach commercial maturity.

Thin-film photovoltaic technologies introduce different dependencies. Cadmium telluride (CdTe) modules rely on tellurium, a rare byproduct of copper refining. Copper indium gallium selenide (CIGS) modules require indium and gallium—materials also used in semiconductors, LEDs, and advanced optics. Although thin-film technologies represent a smaller market share than crystalline silicon, they are strategically significant because they offer complementary advantages such as high performance in diffuse light or reduced reliance on silicon. Yet the byproduct nature of their key materials means supply cannot scale rapidly, limiting their global potential unless new extraction technologies emerge.

Beyond wind and solar hardware, the energy transition demands massive upgrades to electrical grids. Without expanded transmission lines, transformers, converters, and power electronics, renewable energy cannot reach end users. Grid infrastructure is heavily dependent on copper, aluminum, electrical steel, and transformer-grade components. Copper, in particular, is a bottleneck. Renewable-heavy energy systems use far more copper than fossil-fuel-based systems due to the need for extensive transmission and distributed storage. Declining ore grades, long permitting times, and insufficient investment in new copper mines pose a potential threat to global electrification goals.

Power electronics—essential to solar inverters, wind turbine converters, and grid stability—require specialty materials such as silicon carbide and gallium nitride. These wide-bandgap semiconductors enable high-efficiency energy conversion, reduced system losses, and improved thermal management. However, their supply chains depend on high-purity materials, complex crystal-growth processes, and a handful of producers. Silicon carbide substrates, for example, are produced by only a few companies worldwide, and scaling production is technically challenging. The availability of these semiconductor materials may limit how fast renewable energy systems can expand.

Energy storage adds yet another dimension. Renewable energy systems increasingly depend on batteries to address intermittency. Lithium-ion batteries dominate storage markets, though alternative chemistries such as sodium-ion, zinc-based systems, and vanadium-flow batteries are emerging. Lithium, nickel, cobalt, graphite, and manganese shape the economics and availability of utility-scale storage. Vanadium-flow batteries rely on vanadium pentoxide, a material with its own supply constraints. Each storage technology imposes different material requirements, and in aggregate they increase global demand for a suite of critical minerals that are already under pressure from electric vehicle growth.

Hydrogen, frequently cited as a complementary energy transition technology, introduces an additional set of material dependencies. Electrolyzers rely on iridium, platinum, and rare earth catalysts. Solid oxide electrolyzers require high-performance ceramics and specialty alloys. Hydrogen storage tanks rely on carbon fiber, while fuel cells require platinum group metals. The rapid acceleration of hydrogen strategies in Europe, the United States, Japan, and South Korea could strain markets for iridium and platinum, which are already limited by low supply elasticity.

Climate change itself poses significant risks to renewable energy supply chains. Extreme weather disrupts shipping routes used to transport large wind turbine blades or heavy solar modules. Drought has reduced the capacity of the Panama Canal, increasing shipping delays for photovoltaic components. Extreme heat threatens grid reliability and can reduce the efficiency of solar farms. At the same time, renewable projects increasingly face local resistance due to concerns about land use, biodiversity, and environmental impact. Projects may be delayed not by technical limits but by social and regulatory friction.

Logistics also constrain deployment. Wind turbines are enormous structures requiring specialized vessels, ports, cranes, and transportation corridors. Offshore wind relies on installation ships that are in short supply globally. Solar panels travel through long maritime supply chains vulnerable to chokepoints in the South China Sea, Strait of Malacca, and Suez Canal. Any disruption along these routes can delay renewable deployment across entire regions.

The push for domestic renewable manufacturing is reshaping global supply chains. The United States is investing in polysilicon revival, module assembly, and inverter production. Europe is seeking to rebuild solar manufacturing capacity through subsidies and industrial partnerships. India aims to become a major module producer and is investing in upstream capacity. Yet scaling domestic manufacturing requires secure access to raw materials, which these regions often lack. Rebuilding manufacturing without diversifying mineral supply chains risks moving bottlenecks upstream rather than eliminating them.

Recycling will become increasingly important for wind and solar materials. Early models of solar modules are beginning to reach end of life, creating challenges—and opportunities—for recovering silver, silicon, and aluminum. Wind turbine blades, made from composites, are more difficult to recycle, though new processes are under development. Rare earth magnets from wind turbines represent a potentially valuable source of secondary supply if recycling systems become mature. Over the next two decades, recycling could become a stabilizing force in material markets, but only if industry standards, regulatory frameworks, and economic incentives align.

Wind and solar technologies are often described as unlimited because their fuel—wind and sunlight—is renewable. But the materials required to capture that energy are not unlimited. They are finite, difficult to substitute, and embedded in complex geopolitical systems. The success of the energy transition will depend not only on deploying more turbines and panels but on strengthening the supply chains that feed them, diversifying material sources, and developing technologies that reduce dependence on constrained resources.

Semiconductors represent one of the most technologically demanding and geopolitically sensitive domains of the modern economy. Their production rests on a narrow set of materials, extreme precision, and highly concentrated industrial capabilities. In this space, material constraints translate directly into digital power and strategic advantage.

CHAPTER 20 – SEMICONDUCTORS AND THE MINERAL BACKBONE OF COMPUTING

Modern computing owes its existence to materials. From the most advanced artificial intelligence accelerators to the simplest microcontroller, every semiconductor device represents a triumph of materials science, chemical purity, and global supply chains that concentrate critical minerals from around the world. Semiconductors are not abstract technologies but physical objects crafted from some of the rarest, most refined, and most strategically important materials on the planet. Understanding the mineral foundations of computing is essential to understanding the digital economy itself—and the vulnerabilities that threaten it.

At the core of all semiconductor devices lies silicon, the fundamental substrate upon which the global computing infrastructure is built. Yet silicon alone cannot meet the demands of modern electronics. To function effectively, it must be doped with precisely controlled amounts of elements such as phosphorus, boron, and arsenic, each of which alters its electrical properties. These dopants must be introduced in quantities so small that they are measured in parts per billion or even trillion, requiring ultra-high purity gas and chemical supply chains. The production of these chemicals depends on stable access to specialty minerals, industrial gases, and chemical-grade transformation facilities, many of which are concentrated in only a few countries.

As semiconductor devices have advanced, the materials required to manufacture them have diversified dramatically. Whereas early chips relied primarily on silicon, aluminum, and a few simple dopants, today's chips use cobalt and tungsten for contacts and interconnects, rare earth oxides for polishing and dielectric layers, copper for wiring, tantalum for barrier layers, indium and gallium for compound semiconductors, and specialty gases such as neon for lithography. Each of these materials introduces new supply chain dependencies.

High-performance compound semiconductors represent one of the clearest examples of critical material dependence. Gallium arsenide (GaAs) is essential for radio-frequency amplifiers used in mobile phones, radar systems, and high-speed communication devices. Gallium nitride (GaN) has become indispensable for high-power electronics, 5G base stations, and emerging energy-efficient power conversion systems. Indium phosphide (InP) powers high-speed optical communication chips, forming the backbone of global fiber-optic networks. All of these depend on gallium and indium—elements not traded in large commodity markets and produced largely as byproducts of aluminum and zinc refining. As global demand for semiconductors grows, the supply of these niche metals becomes increasingly tight, constrained not by gallium or indium markets themselves but by the economics of aluminum and zinc production.

Further up the chain, the transition to extreme ultraviolet (EUV) lithography has introduced another layer of material dependence. EUV lithography requires mirrors made of molybdenum and silicon, and relies on tin droplets vaporized by lasers to generate EUV light. This is a highly specialized system dominated by a single manufacturer and dependent on materials that require precise, high-purity supply. Any disruption in the availability of tin, molybdenum, or the gases required to operate EUV systems can delay chip production across the globe. The fragility of this supply chain became evident when geopolitical shocks disrupted industrial gas supply, highlighting how semiconductor production depends not only on rare elements but also on reliable access to neon, krypton, and xenon for laser-based manufacturing tools.

As chips increase in complexity, packaging materials have become equally essential. High-performance chips used for artificial intelligence, data centers, and high-end consumer electronics rely on substrates, bonding wires, and interposer structures that incorporate gold, palladium, rare earth oxides, specialized glass, and advanced ceramics. Gold is used extensively for plating electrical contacts, improving conductivity and corrosion resistance. Palladium, with its unique chemical properties, appears in bonding wires and component plating. Rare earth oxides are used in ceramic substrates that must withstand heat, electrical stress, and intense processing conditions. These materials may appear only in small quantities per chip, but multiplied across hundreds of millions of devices, they represent significant global demand.

The connection between semiconductors and critical materials becomes even more visible when examining the power delivery systems in high-performance computing. AI accelerators and graphics processors draw enormous currents, requiring advanced inductors and capacitors made from nickel, iron, tantalum, and specialized ceramics. Tantalum capacitors are particularly important for stabilizing power at the nanosecond scale, and their production depends on high-purity tantalum powders derived from coltan ore. Tantalum supply chains are historically fragile, associated with conflict zones and limited refining capacity. Any disruption can reverberate across the computing sector, affecting devices from data center servers to medical equipment.

The rise of artificial intelligence and cloud computing has intensified these material dependencies. Data centers require thousands of processors, each containing copper wiring, gold plating, rare earth magnets, and gallium-based semiconductor devices. High-bandwidth memory modules rely on materials such as gallium and germanium, and the cooling systems required to support dense compute clusters depend on fluorinated fluids derived from fluorspar. Even the electricity infrastructure feeding these facilities requires copper, aluminum, and advanced power semiconductors, creating an ecosystem in which data infrastructure is intimately tied to the mineral economy.

Optical communication systems—the hidden circulatory system of the global internet—depend on materials even more exotic. Fiber-optic cables rely on germanium doped into silica cores, enabling the precise refractive index properties required for long-distance signal transmission. Erbium-doped fiber amplifiers, critical for boosting signals across continents and undersea cables, depend on rare earths produced primarily in China. Without erbium, the modern telecommunications network would collapse. Meanwhile, the lasers that power fiber networks require indium, gallium, and phosphorus, linking digital communication directly to the availability of critical minor metals.

Semiconductors are also central to emerging technologies such as quantum computing, where material dependencies become even more specialized. Superconducting qubits rely on niobium and aluminum deposited on ultra-pure sapphire or silicon substrates, cooled using helium-3 and helium-4 isotopes. Ion trap quantum computers use ytterbium or barium ions manipulated with lasers made from gallium-based semiconductors. Even the most experimental quantum technologies depend on niche materials sourced from fragile global supply chains. These markets are currently small, but their strategic implications are enormous.

The geopolitical implications of these material dependencies are profound. China's dominance in gallium, germanium, graphite, and rare earth refining gives it leverage over semiconductor supply chains. Its 2023 export controls on gallium and germanium demonstrated how a small disruption in a niche metal can ripple across global chip production. The message was clear: if geopolitical tensions escalate, the global semiconductor ecosystem is vulnerable. The United States, Europe, Japan, and South Korea have responded by accelerating efforts to diversify sources, build strategic reserves, and subsidize domestic semiconductor production. Yet no amount of capital can rapidly replicate the refining infrastructure, technological know-how, and supply chain integration that China has developed over decades.

As chip manufacturing expands to new regions, the demand for critical minerals will grow even faster. Semiconductor fabs require enormous quantities of ultrapure water, industrial gases, high-purity chemicals, and specialized metals. Building new fabs in the United States, Europe, or elsewhere will increase demand for neon, helium, palladium, and specialty metals, tightening global supply even further. The very success of semiconductor diversification efforts will intensify pressure on material supply chains already struggling to meet demand.

Recycling offers a partial solution. Scrap from semiconductor fabrication, electronic waste, and end-of-life devices can provide secondary sources of gold, palladium, copper, rare earths, and other valuable materials. However, recycling semiconductors is technologically challenging due to the tiny quantities of materials embedded within components and the complexity of separating them. While recycling will help reduce long-term material demand, it cannot replace primary supply in the near future.

As semiconductors continue to evolve, the mineral backbone of computing will become even more complex. New materials will emerge, from two-dimensional semiconductors such as molybdenum disulfide to ultra-wide-bandgap materials such as gallium oxide. Each innovation brings new dependencies and new vulnerabilities. The challenge for governments and industries will be to secure these materials without stifling technological progress.

Semiconductors are the brain of the modern world, but their lifeblood flows through global mineral supply chains. To secure the future of computing, nations must invest in diversified supply, resilient refining capacity, and innovation in materials science.

CHAPTER 21 – DEFENSE, AEROSPACE, AND NATIONAL SECURITY SYSTEMS

Defense and aerospace technologies sit at the apex of material intensity. These sectors do not simply consume critical minerals—they depend on them for mission-critical performance, technological superiority, and national security readiness. In military platforms, materials enable capabilities that no amount of software, training, or tactics can replace. Stealth aircraft require specialty coatings and alloys. Missiles depend on rare earth sensors and actuators. Satellites rely on high-purity compounds and radiation-resistant semiconductors. Even the logistics and communication networks that bind modern militaries together rely on mineral-based technologies across every node. Without secure access to these materials, national security systems face vulnerabilities that cannot be patched digitally or solved through diplomacy alone.

Modern defense platforms rely on materials that deliver exceptional strength-to-weight ratios, heat resistance, magnetic performance, optical precision, and electrical properties. These requirements exceed those of commercial industries and often push materials to the limits of physics. As a result, defense supply chains depend heavily on a narrow set of critical minerals, many of which come from limited or politically sensitive geographies.

Rare earths play an indispensable role in defense systems. Neodymium, praseodymium, and dysprosium are essential for high-performance permanent magnets used in guidance systems, actuators, radar platforms, missile steering mechanisms, aircraft control surfaces, and sonar systems. Samarium-cobalt magnets, valued for their temperature stability and resistance to demagnetization, appear in precision-guided weapons, jet engines, and satellite components. Heavy rare earths such as terbium and dysprosium are irreplaceable in systems that must perform reliably under extreme temperature and stress conditions—conditions typical in missile defense, electronic warfare, and naval applications.

Night-vision technologies rely on germanium and indium, two materials also used in optics, infrared detectors, thermal imaging scopes, and missile seekers. These materials absorb and transmit specific infrared wavelengths with exquisite precision, enabling surveillance, targeting, and intelligence capabilities that define modern military superiority. Without germanium lenses, night operations, perimeter security, and reconnaissance missions lose critical effectiveness.

Gallium-based materials appear throughout advanced defense electronics. Gallium nitride (GaN) powers high-frequency, high-efficiency radar systems such as AESA (Active Electronically Scanned Array) radars, which provide superior detection range, tracking precision, and resistance to jamming. GaN's ability to operate at high power densities and temperatures makes it ideal for electronic warfare (EW) platforms and satellite communication systems. Gallium arsenide (GaAs) underpins radio-frequency weapon systems, secure satellite links, and battlefield communication networks. These semiconductors deliver performance that silicon alone cannot match, making gallium one of the most strategically sensitive minerals in modern defense.

Aerospace systems impose their own extreme requirements. Jet engines operate under intense heat and mechanical stress, relying on nickel-based superalloys strengthened with chromium, cobalt, niobium, tantalum, and rhenium. These alloys enable turbine blades to withstand temperatures that exceed the melting point of many metals. Rhenium is especially critical; even small amounts dramatically increase creep resistance in turbine blades. The United States relies on limited global sources of rhenium, making it a metal of acute strategic importance. Any constraint in its supply directly affects the production and maintenance of high-performance aircraft engines.

Titanium is another cornerstone of aerospace materials. Its strength-to-weight ratio, corrosion resistance, and biocompatibility make it essential for aircraft frames, armor plating, naval vessels, and missiles. Titanium alloys reduce structural weight while maintaining integrity under stress, contributing to increased fuel efficiency and maneuverability. However, titanium sponge production is concentrated in only a few countries, and geopolitical shocks have demonstrated how vulnerable aerospace supply chains are to disruptions in titanium output.

Defense electronics rely on tantalum for high-reliability capacitors used in avionics, missile systems, secure communication devices, and satellite hardware. Tantalum capacitors withstand extreme vibration, temperature, and radiation environments. They enable stable power delivery in mission-critical electronics where failure is unacceptable. Tantalum supply chains have long been associated with geopolitical complexity due to concentration in central Africa, limited refining capacity, and governance challenges. This creates strategic vulnerabilities that ripple through defense procurement systems.

Satellite systems depend on a wide range of materials beyond gallium, germanium, and rare earths. Solar panels for satellites use gallium arsenide cells for high conversion efficiency and radiation resistance. Reaction wheels rely on rare earth magnets. Radiation-hardened microelectronics require specialty silicon, germanium, and insulating oxides. Structural components use aluminum-lithium alloys, titanium, and advanced carbon composites. Even rocket propellants require specialty materials such as aluminum powder and beryllium for ignition systems. Beryllium, valued for its stiffness and thermal stability, is also used in sensors, optics, and guidance systems. Yet beryllium mining and refining are limited and tightly controlled due to toxicity and strategic importance.

A broader view of national security reveals additional mineral dependencies. Cybersecurity hardware, secure servers, and cryptographic devices rely on high-purity silicon, rare earth permanent magnets, gold wiring, and tantalum capacitors. Drone systems require lithium-ion batteries, rare earth magnets, and lightweight composite materials. Undersea cables, critical for global military communication, rely on copper conductors and erbium-doped fiber amplifiers. Naval platforms require copper-nickel alloys for corrosion resistance, manganese for steel, and titanium for pressure-resistant submarine hulls.

Logistics and supply support systems—often overlooked—carry their own dependencies. Military vehicles transitioning to hybrid or electric powertrains increase reliance on lithium, nickel, copper, rare earths, graphite, and cobalt. Portable energy storage for field operations requires lithium-ion batteries and, increasingly, fuel cells dependent on platinum-group metals. These same materials also appear in emergency communication systems, radar installations, and mobile command centers.

The vulnerabilities created by these material dependencies are significant. Many defense-critical minerals come from countries with political instability, weak environmental governance, or strategic rivalry with major powers. China's dominance in rare earth processing and magnet manufacturing poses clear risks for nations whose defense industries rely on steady supplies. Export controls or disruptions could impair missile guidance production, reduce radar performance, or delay aircraft maintenance. Similarly, reliance on Congo for cobalt or South Africa for platinum group metals introduces geopolitical risk beyond the control of consuming nations.

Supply chain fragility in defense is amplified by the sector's need for absolute reliability. Unlike commercial industries, defense programs cannot simply switch materials or redesign components quickly. Military systems often require decades-long supply commitments, strict quality standards, and tightly controlled production environments. A shortage of a single material—such as dysprosium, rhenium, or gallium—can halt production lines, delay procurement cycles, and undermine operational readiness.

Because of these risks, many countries maintain stockpiles of defense-critical materials. The United States' National Defense Stockpile includes rare earths, beryllium, tungsten, cobalt, and other strategic minerals. Japan has similar programs for rare earths and semiconductors. These stockpiles provide short-term resilience but do not solve the structural vulnerabilities created by concentrated supply chains. Their effectiveness depends on accurate forecasting of military technology needs, stable procurement planning, and geopolitical foresight.

Allied collaboration is becoming increasingly important. The United States, Europe, Australia, Japan, and South Korea are coordinating to diversify supply chains, develop new processing capacity, and reduce reliance on adversarial nations. Joint investments in rare earth separation, magnet production, new refining technologies, and co-development of military components aim to build collective resilience. Defense procurement programs increasingly require material origin traceability, ethical sourcing, and supplier diversification.

Innovation also plays a central role in risk reduction. Research into rare-earth-free motors, gallium alternatives, next-generation superalloys, and new optical materials aims to reduce dependence on vulnerable supply chains. Additive manufacturing (3D printing) is creating new ways to produce components with optimized material usage, reducing waste and enabling new alloy designs. Yet substitution has limits. Many critical minerals are unmatched in their functional properties, meaning that true independence from them is technologically unrealistic in the near term.

CHAPTER 22 – ELECTRONICS, OPTICS, AND MEDICAL TECHNOLOGIES

Modern civilization is built on electronics. From smartphones to satellites, from hospital imaging systems to the vast network of devices forming the Internet of Things, contemporary life depends on technologies that require an extraordinary—and often invisible—array of critical materials. These sectors represent the quiet but pervasive infrastructure of the digital, interconnected world. Unlike defense or aerospace, electronic and medical technologies operate at massive scale, embedding rare and strategically important minerals into billions of devices each year. The result is a vast, constantly moving ecosystem dependent on materials whose supply chains remain fragile, concentrated, and vulnerable to disruption.

Consumer electronics form one of the largest and most mineral-intensive technological categories. Smartphones alone contain dozens of critical materials in extremely small but fundamentally irreplaceable quantities. Tantalum capacitors regulate power flow in compact circuits; gallium and indium form the compound semiconductors used in radio-frequency amplifiers and touchscreens; neodymium magnets drive haptic motors, speakers, and vibration modules; cobalt and lithium power the battery; and gold, silver, and palladium are used in connectors and wiring. The complexity and density of these materials reflect the continuous push toward miniaturization, multifunctionality, and high-performance consumer devices.

As the global population adopts more smartphones, wearables, laptops, tablets, and smart appliances, demand for these materials grows insatiably. Electronics are produced in the hundreds of millions annually, meaning even tiny quantities of critical minerals accumulate into significant global usage. What makes electronics particularly challenging is that most of these materials are extremely difficult to recover at end of life. Devices are compact, intricately assembled, and often discarded long before recycling systems can capture their mineral content. This results in a constant need for primary material extraction, deepening dependence on vulnerable supply chains.

Optoelectronics—the technologies that convert electrical signals into light and vice versa—underscore another set of material dependencies. Light-emitting diodes (LEDs) depend on gallium, indium, and rare earth phosphors such as europium and terbium. LEDs have become the global standard in lighting, displays, automotive electronics, and industrial systems due to their energy efficiency, long lifespan, and compactness. However, their production relies on materials produced in small quantities and concentrated in a few countries. For example, gallium is primarily recovered as a byproduct of aluminum refining, and indium depends on zinc smelting. These byproduct dynamics limit scalability and increase market volatility.

Display technologies further reflect the increasingly complex mineral foundation of the electronics sector. Liquid crystal displays (LCDs) require indium-tin oxide (ITO) for transparent conductive coatings. Organic LEDs (OLEDs), used in premium televisions and smartphones, require specialty organic compounds synthesized through rare catalysts, as well as precision-deposited thin films that depend on germanium and other high-purity inputs. Quantum dot displays rely on indium phosphide or cadmium-based compounds to deliver high color accuracy. Each of these display technologies depends on specific material chains with limited flexibility.

Medical technologies introduce even higher stakes. Unlike consumer electronics, which rely on materials for performance and efficiency, medical devices depend on them for reliability, safety, and diagnostic accuracy. Magnetic resonance imaging (MRI) machines depend heavily on superconducting magnets made from niobium-titanium alloys and cooled with helium—especially helium-3, one of the rarest and most geopolitically sensitive isotopes in the world. Helium scarcity has already disrupted MRI operations in some regions, demonstrating how medical infrastructure can be directly impacted by strains in global material supply.

X-ray and CT imaging systems rely on tungsten for radiation shielding and molybdenum for x-ray tube components. Ultrasound devices require piezoelectric crystals, often composed of lead zirconate titanate, which depend on high-purity zirconium and titanium. Medical lasers use rare earth dopants such as erbium or holmium. Even simple medical tools like scalpels, orthopedic implants, or dental instruments rely on high-performance alloys made from cobalt, chromium, titanium, and nickel.

In diagnostics, optoelectronic sensors incorporate compound semiconductors and rare earth phosphors to detect biological signals. Point-of-care diagnostic equipment uses gallium-based components for rapid signal processing and conversion. Laboratory instruments depend on high-purity glass derived from silicon and boron, as well as precise ceramic components made from aluminum oxide or zirconium oxide. The rise of personalized medicine—built on genetic analysis and continuous data monitoring—will only increase this reliance on advanced materials, since smaller and more precise sensors require materials with exceptional electronic, optical, and chemical performance.

Wearable medical technologies sit at the intersection of consumer electronics and healthcare. Smartwatches, glucose monitors, heart-rate analyzers, and portable ECG devices embed sensors made from rare metals, batteries requiring lithium and cobalt, and communication modules dependent on gallium-based radio-frequency chips. These tools are becoming essential for preventive medicine, aging populations, and digital health systems. As adoption expands, so does demand for the underlying materials and the need to secure their supply chains.

Telecommunications infrastructure—which supports both consumer electronics and medical systems—adds another layer of material dependencies. Fiber-optic networks rely on germanium-doped silica fibers and erbium-doped amplifiers. 5G antennas depend heavily on gallium nitride semiconductors for high-frequency, high-power performance. Base stations require rare earth magnets, copper coils, high-purity aluminum alloys, and advanced ceramics. The rollout of 5G and the future development of 6G networks will exponentially increase demand for these minerals, linking the entire digital ecosystem to supply chains often concentrated in China and Southeast Asia.

Even the automated manufacturing systems used to produce electronics and medical devices rely on critical materials. Industrial robots use servo motors with neodymium magnets. Precision machining tools require tungsten carbide inserts. Cleanroom equipment depends on high-purity stainless steel, aluminum, and specialty plastics. Semiconductor manufacturing plants—the backbone of the electronics and medical technology sectors—consume rare gases, ultrapure chemicals, gallium, indium, germanium, and various metal targets for sputtering and deposition. Any disruption in these supply chains can halt production across entire industries.

Energy storage technologies underpin many medical and electronic devices. Implantable medical devices rely on high-reliability batteries with chemistries specifically designed to minimize risk and maximize longevity. These batteries require high-purity lithium, manganese, titanium, and specialty electrolytes. Backup power systems for hospitals depend on large-scale batteries containing lithium, nickel, cobalt, and graphite. As hospitals transition toward more digital and distributed systems, their reliance on battery-based energy security grows proportionately.

Material vulnerability is a recurring theme across electronics, optics, and medical technologies. Many of the materials they rely on—gallium, germanium, indium, rare earths, helium, tantalum—are produced in geographically concentrated regions. China dominates rare earth refining, magnet production, gallium and germanium output, and the processing of many electronic materials. The Democratic Republic of Congo supplies most cobalt. Russia has historically played a major role in palladium. South Africa dominates iridium and platinum group metals. These concentrations create risks that ripple across digital and medical sectors.

Recycling, though promising, remains technologically difficult in these domains. Electronics contain tiny quantities of valuable materials dispersed across complex assemblies. Recovering these materials is labor-intensive and often uneconomical. Medical devices pose additional challenges due to contamination risks and strict sterilization requirements. Nevertheless, advances in urban mining and automated disassembly technologies are beginning to show promise. Future recycling improvements could relieve pressure on primary extraction, but meaningful impact will take years to materialize.

Innovation offers partial relief. Researchers are developing rare-earth-free permanent magnets, indium-free transparent conductors, gallium-free RF chips, and alternative materials for optical and medical applications. Battery developers are exploring chemistries with reduced cobalt or entirely different architectures. Display manufacturers are experimenting with carbon-based materials and perovskite semiconductors. Yet these innovations often face performance limitations or new dependencies on equally rare materials. Substituting one strategic mineral for another is rarely a clean solution—it simply shifts demand across the periodic table.

CHAPTER 23 – INDUSTRIAL CATALYSTS, ALLOYS, AND ADVANCED MANUFACTURING

Every industrial system—whether producing fuel, chemicals, machinery, or electronics—depends on a hidden architecture of catalysts, alloys, and high-performance materials. These materials rarely enter public debate, yet they underpin the productivity and efficiency of global manufacturing. They accelerate chemical reactions, strengthen metals, improve thermal resistance, and enable high-precision fabrication. Without them, modern industry would collapse into inefficiency, higher costs, and technological stagnation. Their importance is amplified by the fact that many rely on the same critical minerals that shape energy, electronics, and defense supply chains. As industries transition toward cleaner technologies, these foundational materials are becoming even more central, driving demand for a diverse suite of critical elements.

Catalysts form one of the most strategically important but least visible classes of industrial materials. In chemical production, petroleum refining, emissions control, and emerging clean-energy technologies, catalysts determine reaction speed, efficiency, selectivity, and environmental performance. Platinum group metals—platinum, palladium, rhodium, ruthenium, iridium, and osmium—are the crown jewels of catalytic science. Their unique electron structures allow them to moderate chemical reactions under extreme conditions. Palladium and platinum dominate automotive catalytic converters, transforming harmful exhaust gases into less harmful emissions. Rhodium, among the rarest and most expensive metals, enhances these reactions further, enabling compliance with stringent environmental standards.

Iridium and platinum are essential to hydrogen production. Proton-exchange membrane (PEM) electrolyzers require iridium for oxygen evolution and platinum for hydrogen evolution. As nations expand their hydrogen economies, demand for these materials could outpace supply, especially because iridium production is limited to small byproduct streams from platinum mining. Industrial catalysts therefore sit at the intersection of environmental regulation, energy transition strategies, and material scarcity.

Beyond precious metals, catalytic materials such as cerium oxide, lanthanum-based catalysts, vanadium, nickel, cobalt, and molybdenum play major roles in petroleum refining and chemical synthesis. Worldwide, refineries depend on these catalysts to crack hydrocarbons, remove impurities, and produce fuels with precise performance characteristics. Even as fossil fuel dependence declines over time, the petrochemical sector remains essential for plastics, solvents, lubricants, and countless industrial products—ensuring that catalytic materials will remain critical for decades.

Industrial alloys form another backbone of manufacturing. High-temperature alloys used in turbines, engines, furnaces, and reactors incorporate nickel, chromium, cobalt, tungsten, molybdenum, vanadium, and niobium. These elements strengthen metal structures, improve corrosion and heat resistance, and enable operations in extreme environments. Nickel-based superalloys, essential for jet engines, gas turbines, and industrial machinery, cannot function without cobalt, chromium, and rhenium. Small additions of rhenium dramatically improve creep resistance, making it indispensable for advanced aerospace engines. Yet rhenium is scarce, difficult to recover, and produced in limited quantities as a byproduct of copper mining.

Specialty steels further illustrate the importance of critical materials. Manganese is essential for steelmaking. Vanadium strengthens steel for tools, pipelines, and rebar. Chromium enables stainless steel. Tungsten carbide tools—made from tungsten and cobalt—form the cutting edges of machining equipment central to modern manufacturing. These materials allow factories to shape, cut, and form metals with precision and durability. Without tungsten carbide, industrial machining would slow dramatically, increasing costs and reducing global manufacturing efficiency.

Titanium alloys provide high strength-to-weight ratios and corrosion resistance, making them essential not only for aerospace but for industrial heat exchangers, chemical processing equipment, and desalination plants. Zirconium alloys are indispensable in nuclear reactors, thanks to their low neutron absorption and high corrosion resistance. Hafnium, often extracted alongside zirconium, is used in control rods for nuclear power plants and is also a crucial material for high-k dielectric layers in semiconductor manufacturing. These overlapping uses illustrate how industrial and technological sectors compete for the same scarce materials.

Advanced manufacturing technologies amplify this dependence. Additive manufacturing, commonly known as 3D printing, has revolutionized prototyping, tooling, and increasingly end-use part production. High-performance metal powders used in 3D printing rely on titanium, nickel, aluminum, and specialty alloys. These powders must maintain extremely tight tolerances in composition and granularity. As industries adopt additive manufacturing for aerospace components, medical implants, and high-precision tools, demand for these powders—and the critical minerals embedded within—continues to rise.

Ceramics and advanced composites represent another frontier. Industrial ceramics used in furnaces, cutting tools, sensors, and electronics rely on rare earth oxides, zirconium, aluminum oxide, silicon carbide, and boron carbide. These materials withstand temperature, abrasion, and chemical stress that destroy conventional metals. Silicon carbide, already vital for power electronics, also forms kiln linings, abrasives, and high-strength components used in energy-intensive industries. Boron carbide and silicon nitride composites appear in armor, industrial nozzles, and precision bearings. Each of these relies on supply chains intertwined with minerals that carry significant geopolitical risk.

Glass and optical materials also deserve attention. Specialty glass used in industrial lasers, high-temperature environments, and telecommunications relies on rare earth dopants, high-purity silica, and boron-based compounds. The same materials that enable fiber optics, sensors, and GPS systems also appear in manufacturing machinery, creating dual-use dependencies that integrate industrial and technological sectors tightly.

The transition toward cleaner industrial processes introduces new material challenges. Electrified furnaces, hydrogen-ready steelmaking, carbon capture systems, and new catalytic pathways require materials with advanced catalytic and thermal properties. Electrolyzer components depend not only on platinum and iridium but also on polymer membranes, nickel foams, and rare-earth coatings. Catalysts optimized for carbon dioxide conversion or green ammonia production require novel combinations of ruthenium, cobalt, and advanced nanostructured materials. These innovations accelerate the material intensity of the industrial transition.

Automation and robotics rely heavily on critical materials as well. High-precision motors use rare earth magnets. Industrial sensors use germanium and indium. Power systems rely on silicon carbide semiconductors. Modern automated factories depend on a supply chain of minerals that extends across every continent and is subject to geopolitical and logistical pressures. If material supply falters, manufacturing systems slow or stop—not because machines break, but because their critical components cannot be replaced.

As industries digitalize, their reliance on these materials deepens. Machine-learning-driven process optimization depends on high-performance computing, which itself depends on semiconductors, gallium, germanium, copper, and gold. Industrial networks depend on fiber optics doped with rare earths. Smart manufacturing ecosystems rely on sensors, actuators, and power electronics built from rare materials. The digital transformation of manufacturing increases its exposure to the vulnerabilities already seen in the electronics and semiconductor sectors.

Recycling offers potential relief but remains challenging. Many industrial catalysts can be reclaimed, especially platinum group metals. High-performance alloys can be recycled, though separating critical materials remains energy-intensive. Rare earth magnets can be recovered from industrial motors and machinery, but magnet recycling remains in its infancy. Tungsten carbide recycling is relatively mature and contributes meaningfully to global tungsten supply. However, composite materials, ceramics, and high-precision alloys are much more difficult to recycle at scale. Without substantial innovation, recycling will offset only part of industrial mineral demand.

Innovation will shape the future of industrial materials. Researchers are developing cobalt-free or low-cobalt alloys, rare-earth-free magnets, advanced catalysts that use less precious metal, and high-temperature alloys based on abundant elements. Nanomaterial catalysts promise improved efficiency with lower material intensity. Ceramic composites designed for additive manufacturing could reduce reliance on scarce metals. Yet new materials often introduce new dependencies, requiring their own specialized supply chains.

Iindustrial catalysts, alloys, and advanced materials form the skeleton and circulatory system of global manufacturing. They are essential to energy production, chemical synthesis, transportation, construction, and digital technology. Their importance is often overshadowed by downstream products, but no industrial system can function without them. As global demand increases—and as geopolitical tensions reshape access to critical minerals—industrial supply chains will face mounting pressure.

Pressure on primary supply is pushing material strategies beyond extraction. Urban mining, recycling, and secondary materials are emerging as mechanisms to reduce vulnerability and extend resource lifecycles. Their effectiveness will shape whether material security can evolve toward greater resilience and sustainability.

CHAPTER 24 – URBAN MINING, RECYCLING, AND SECONDARY MATERIALS

The shift toward a sustainable, low-carbon global economy is reshaping not only how societies produce and use energy, but how they source the materials required to build the infrastructure of the twenty-first century. While traditional mining will remain indispensable for decades, the world is beginning to recognize the immense value contained in materials already in circulation—embedded in electronics, vehicles, turbines, industrial machinery, and countless other technologies. Urban mining and recycling, once considered niche or supplementary, are now emerging as strategic pillars of resource security. They offer a pathway to reduce dependence on geopolitically concentrated supply chains, mitigate environmental impact, and create a more circular, resilient materials system.

Urban mining refers to the systematic recovery of valuable materials from waste streams, buildings, products, and infrastructure in urban environments. The concept rests on a simple but powerful observation: cities contain more concentrated deposits of metals, minerals, and alloys than many natural ore bodies. A tonne of discarded smartphones contains far more gold than a tonne of mined ore; retired wind turbines contain large quantities of rare earth magnets; end-of-life electric vehicles hold lithium, cobalt, nickel, and copper; and obsolete industrial machinery is rich in tungsten, steel alloys, and permanent magnets. These "above-ground mines" are growing rapidly as global consumption increases and product lifespans end.

Recycling, the technological backbone of urban mining, transforms waste into secondary raw materials. When executed effectively, recycling reduces the need for primary extraction, lowers carbon intensity, and provides countries with domestic sources of critical minerals they might otherwise need to import. Yet recycling is far from simple. Every technology—from a smartphone battery to a wind turbine blade—has its own chemical and structural complexities that determine how easily materials can be recovered.

Lithium-ion battery recycling is among the most strategically important areas of urban mining. As electric vehicles and stationary storage systems proliferate, end-of-life batteries will become a massive secondary resource for lithium, cobalt, nickel, manganese, and graphite. Modern recycling processes use pyrometallurgical, hydrometallurgical, or hybrid techniques to extract metals from shredded battery cells. Pyrometallurgy melts batteries at high temperatures, producing a metal-rich "black mass," while hydrometallurgy dissolves metals into solution, enabling more selective separation. Both approaches can achieve high recovery rates for nickel, cobalt, and copper, but recovering lithium and graphite at scale remains challenging.

Battery recycling is not merely an environmental exercise—it is a strategic one. Because the majority of global battery refining capacity is concentrated in China, countries pursuing battery independence view recycling as a way to reduce reliance on foreign supply chains. Over the next decade, recycling is expected to become a significant source of cobalt and nickel, with lithium catching up as technologies improve. The long-term vision is a closed-loop battery ecosystem in which materials from end-of-life packs feed directly back into new cell production.

Permanent magnets represent another high-value target for urban mining. Neodymium-iron-boron magnets, embedded in electric vehicle motors, wind turbines, industrial machinery, and consumer electronics, contain rare earth elements that are difficult to substitute and heavily concentrated in Chinese supply chains. Recycling these magnets typically involves demagnetization, separation from steel components, and chemical processing to recover rare earth oxides. Emerging technologies such as hydrogen decrepitation show promise in simplifying magnet recycling by causing magnets to crumble into powders that can be directly reprocessed.

Despite their potential, rare earth magnet recycling faces major obstacles: magnets are dispersed across countless products, often glued or molded into components, making extraction labor-intensive. Many older devices contain small quantities of rare earths that are difficult to recover economically. Nonetheless, as electric vehicles and wind turbines reach end of life, the volume of available magnets will grow dramatically, improving the economics of magnet recycling and creating opportunities for countries seeking to build domestic rare earth value chains.

Consumer electronics offer another rich resource for urban mining. Smartphones, laptops, servers, and appliances contain gold, silver, palladium, copper, tin, indium, and rare earths. Printed circuit boards are among the most mineral-dense waste streams on earth. Yet e-waste recycling remains a global weakness. Much e-waste is exported to countries lacking adequate recycling infrastructure, where informal processes expose workers to health risks and recover only a fraction of valuable materials. Advanced recycling techniques—such as automated disassembly, laser separation, and bio-leaching—are emerging, but adoption remains limited. To unlock the full potential of e-waste, governments will need to enforce stricter recycling standards, promote producer responsibility, and support investments in advanced processing facilities.

Wind and solar technologies are also becoming part of the recycling landscape. Wind turbine blades, made primarily from fiberglass composites, are notoriously difficult to recycle. Solutions include mechanical grinding, chemical decomposition, and high-temperature processes that recover fibers and resins for reuse. Solar photovoltaics contain silver, aluminum, glass, silicon, and specialty plastics. Crystalline silicon modules are increasingly being recycled using thermal, chemical, or mechanical processes. Silver recovery is particularly important given solar's large share of global silver demand. While solar recycling is in its infancy, the rapid expansion of solar deployment means that recycling infrastructure will become essential in the coming decades.

Industrial machinery and infrastructure provide additional urban mining opportunities. Retired industrial motors contain permanent magnets; machine tools embed tungsten carbide; pipelines incorporate steel alloys with chromium, molybdenum, and vanadium; and decommissioned power plants contain copper, aluminum, and specialty steels. In many advanced economies, industrial dismantling and scrap recycling already provide significant metal supply. However, recovering specialty materials remains technologically challenging and often overlooked in national resource strategies.

Construction and demolition waste—one of the largest waste streams globally—contains steel, copper wiring, aluminum, and sometimes rare earth–containing components from HVAC systems and elevators. While steel is already heavily recycled, recovering other metals from buildings requires more systematic approaches, including improved building design for disassembly and detailed material inventories known as "material passports."

Urban mining's potential is immense, but several barriers must be addressed to realize it fully. First, collection systems are often fragmented or inadequate. Many devices remain in households long after they are obsolete, creating a significant "hibernating stock" of valuable materials. Others are discarded improperly or mixed with general waste. Creating effective take-back programs, deposit systems, and incentives for consumers is essential.

Second, recycling technologies must advance. Many critical minerals are embedded in complex products where chemical or mechanical separation is difficult. High-value metals such as gold and palladium are relatively easy to recover, but rare earths, gallium, indium, and lithium require specialized processing that is not widely available outside a few countries. Breakthroughs in solvent extraction, direct recycling, and electrometallurgical processing could dramatically expand what is economically recoverable.

Third, economic viability is a persistent challenge. Recycling must compete with primary mining, which often enjoys economies of scale, established infrastructure, and—in some regions—lax environmental standards. Recyclers face price volatility and uncertain material flows. Effective policy measures, such as minimum recycled-content requirements, extended producer responsibility laws, and subsidies for advanced recycling facilities, can improve economic conditions for urban mining.

Fourth, environmental considerations must be tightly managed. Poorly regulated recycling can produce toxic emissions, wastewater contamination, and hazardous working conditions. True circularity requires responsible systems that reduce environmental harm, not shift it to different regions.

Despite these challenges, opportunities for urban mining are expanding rapidly. Governments in Europe, Japan, South Korea, and North America are promoting circularity as a strategic pillar of material security. The emergence of battery gigafactories, semiconductor fabs, and renewable energy clusters is creating demand for local, reliable secondary material sources. Companies are developing business models around closed-loop manufacturing, where production scrap and end-of-life products feed directly into new material streams.

Over the long term, urban mining could significantly alter global material flows. It will not eliminate the need for primary mining—especially as developing economies grow and new technologies emerge—but it can reduce pressure on vulnerable supply chains, lower environmental footprint, and strengthen industrial autonomy. As billions of devices, vehicles, and machines reach the end of their useful life, cities will become some of the richest mines on the planet.

Technological innovation is beginning to reshape material demand rather than simply absorb its constraints. Substitution, efficiency gains, and novel material solutions offer pathways to alter dependency profiles across key industries. Whether these advances relieve pressure or introduce new vulnerabilities remains a central strategic question.

CHAPTER 25 – SUBSTITUTION TECHNOLOGIES AND MATERIAL INNOVATION

For decades, the global economy has been shaped by the availability and strategic importance of key minerals. But as demand for critical materials rises faster than supply can adapt, industries and governments are accelerating efforts to reduce dependence on vulnerable supply chains through substitution and material innovation. Substitution is not a single solution; it is a complex strategies, technologies, and design philosophies that attempt to replace scarce, geopolitically concentrated, or environmentally problematic materials with more abundant, stable, or sustainable alternatives. Material innovation extends this logic by developing entirely new classes of materials—structural, electronic, magnetic, catalytic—that deliver equal or superior performance while easing pressure on critical supply chains. Together, these approaches represent some of the most important pathways toward a more resilient and sustainable technological future.

Substitution is not simply about replacing one mineral with another. It requires rethinking the fundamental physics, chemistry, and engineering principles behind products and systems. Some materials resist substitution because their properties are unique or difficult to match; others face partial substitution possibilities depending on performance requirements. In many cases, substitutions shift material demand rather than eliminate it, creating new dependencies that must be considered. The story of substitution is therefore a story of trade-offs—economic, technological, and geopolitical.

Permanent magnets offer one of the most visible and consequential fronts in material substitution. Neodymium-iron-boron magnets, enhanced with dysprosium or terbium, offer unmatched strength-to-weight ratios and efficiency. They are critical for electric vehicle motors, wind turbines, robotics, aerospace systems, and precision industrial machinery. Efforts to reduce reliance on rare earth magnets involve both incremental and radical approaches. Incrementally, researchers are developing magnets that use lower amounts of dysprosium and terbium by optimizing grain boundary diffusion techniques or improving thermal stability through grain engineering. Radically, scientists are exploring rare-earth-free magnet designs based on materials such as manganese-aluminum-carbon, iron-nitrogen, and cobalt-iron alloys. Some of these show promise but remain below the performance thresholds required for EV motors or offshore wind turbines. Substitution in magnets illustrates a broader truth: performance requirements in cutting-edge technologies often leave little margin for compromise.

Battery chemistries represent another major frontier of substitution. The move away from cobalt-rich cathodes reflects both supply concerns and cost pressures. Next-generation lithium-ion batteries increasingly favor nickel-rich chemistries (such as NMC 811) or cobalt-free variants. Lithium iron phosphate (LFP) batteries, which eliminate both nickel and cobalt, have re-emerged as a dominant technology for mass-market electric vehicles due to their stability, long cycle life, and lower cost. The rapid scaling of LFP manufacturing, especially in China, demonstrates how substitution can reshape global markets.

Beyond lithium-ion, alternative battery technologies promise deeper substitution. Sodium-ion batteries eliminate lithium and use abundant sodium instead. Although energy density remains lower than lithium-ion, sodium-ion systems are well-suited for stationary storage and low-cost electric vehicles. Zinc-based batteries provide safe, low-cost alternatives for grid applications. Vanadium flow batteries, while dependent on vanadium, offer long-duration storage with high recyclability. Silicon anodes, replacing graphite, are emerging in high-performance batteries, though their adoption requires solving challenges related to swelling and cycle degradation. Solid-state batteries, if commercialized at scale, could reconfigure material requirements entirely by enabling new anode and cathode combinations and reducing the need for liquid electrolytes.

In the semiconductor sector, substitution efforts focus on reducing reliance on gallium, indium, germanium, and other minor metals essential for compound semiconductors. Research into silicon photonics seeks to shift certain optical communication functions away from indium phosphide. Organic semiconductors show promise for flexible electronics, potentially reducing reliance on rare metals in display technologies. Zinc oxide is being explored as an alternative material for transparent conductive layers in place of indium-tin oxide (ITO). While these technologies offer potential diversification, they often face limitations in efficiency, durability, or commercial scalability.

Optical and photonic materials also present opportunities for substitution. Perovskite solar cells, for example, reduce reliance on high-purity silicon and may eventually lower silver intensity by using different contact materials. In displays, carbon-based materials are being investigated as possible replacements for rare metal–dependent components. In lasers and sensors, new dopant combinations and synthetic crystals aim to reduce reliance on terbium, europium, and other rare earths, though replacing their unique optical emissions remains extraordinarily difficult.

Substitution also plays a key role in catalysts and industrial chemistry. Researchers are developing nickel- or iron-based catalysts for hydrogen production, potentially reducing iridium and platinum demand. Copper-based catalysts are being explored for CO_2 conversion processes. Manganese and molybdenum variants are being tested for ammonia synthesis and hydrogen storage. Yet platinum group metals remain exceptionally effective, and their catalytic behavior is difficult to replicate, meaning meaningful substitution will require significant breakthroughs in both computational materials science and industrial chemistry.

In structural materials, substitution is often more straightforward but still consequential. Aluminum alloys substitute for copper in electrical systems, especially in transmission lines where weight and cost matter more than conductivity. Magnesium alloys substitute for aluminum in lightweighting applications. High-strength steels compete with titanium in some aerospace and automotive contexts. Polymer composites can replace steel in certain structural applications, though they introduce their own recycling challenges and rely on petrochemical feedstocks. Substitution in alloys tends to reflect cost-performance optimization rather than strategic dependency, but demand shifts can still impact critical mineral markets.

One of the most promising avenues in substitution is the rise of engineered and metamaterials. These materials achieve properties through structure rather than composition. For example, architected lattices can deliver exceptional strength using abundant materials like carbon, aluminum, or silicon. Photonic crystals can manipulate light with minimal use of rare optical dopants. Magnetic metamaterials may one day mimic the functional performance of rare earth magnets without requiring rare earths themselves. These technologies are still largely experimental, but they illustrate the transformative potential of material science innovation.

Another emerging frontier is bio-based materials. Bio-derived polymers, resins, and composites offer sustainable alternatives to traditional petrochemical materials. Biological systems can produce highly ordered structures at room temperature and atmospheric pressure, presenting opportunities to replicate or even surpass the performance of conventional materials using renewable feedstocks. Importantly, bio-inspired materials may allow engineers to avoid scarce minerals entirely in certain applications, though replacing the electronic or magnetic properties of metals remains extremely challenging.

Digital innovation accelerates progress in material science. Machine learning and high-throughput computational screening allow researchers to explore millions of potential material combinations in silico before synthesizing promising candidates in the lab. This accelerates discovery and reduces the cost of experimentation. Coupled with additive manufacturing, researchers can quickly test and iterate new material formulations in real-world geometries. These tools expand the frontier of substitution technologies, enabling solutions that once seemed impossible.

Despite these advances, substitution faces inherent limitations. Some materials are irreplaceable for certain functions. Rare earths possess magnetic and optical properties unmatched by any known substitute. Copper's conductivity is difficult to rival at scale. Lithium remains uniquely suited for high-energy-density batteries. Helium's cryogenic properties cannot be replicated by any known material. Substitution, therefore, is not a universal solution—it is a strategic tool to reduce risk, reorganize demand, and buy time for innovation.

Moreover, substitution often shifts rather than solves dependency. Replacing cobalt with nickel, for example, increases reliance on nickel supply chains. Moving from silicon to perovskites shifts reliance to lead or other sensitive materials. Substituting indium for zinc oxide improves abundance but can reduce performance. These dynamics illustrate that substitution must be evaluated holistically, considering geopolitical, environmental, and technical implications.

Substitution technologies and material innovation are essential components of the broader strategy for securing critical materials in a world increasingly defined by electrification, digitalization, and sustainability. They complement efforts in recycling, urban mining, and supply chain diversification. Most importantly, they shape the future trajectory of technology itself.

Substitution pathways intersect unevenly with demand for magnets, batteries, and semiconductors. In some cases they soften constraints; in others they shift pressure onto different materials and supply chains. The material landscape ahead will be shaped less by absolute scarcity than by how these technological choices redistribute risk and dependency.

CHAPTER 26 – THE FUTURE OF MAGNET, BATTERY, AND SEMICONDUCTOR MATERIALS

The future of global technology will be shaped not only by innovation in algorithms, electronics, and renewable energy but by the evolution of the materials that make these systems possible. Magnets, batteries, and semiconductors—three foundational pillars of the modern age—are all entering periods of profound transformation. Rising demand, geopolitical pressure, supply chain vulnerabilities, and rapid technological shifts are converging to reshape which materials will dominate the next decades. As nations and industries attempt to secure their technological independence, new material pathways are emerging that will redefine performance, cost, and the balance of global power.

Permanent magnets sit at the heart of electrification. They power electric vehicle motors, robotics, wind turbines, drones, industrial machinery, and nearly all high-performance motion systems. Today, neodymium-iron-boron magnets dominate because they offer unparalleled power density and efficiency, especially when enhanced with dysprosium or terbium for thermal stability. However, the rare earth supply chain remains heavily concentrated, and demand is expected to soar as EV adoption accelerates and offshore wind turbines increase in size. This combination of scarcity and strategic vulnerability is driving the global search for next-generation magnetic materials.

Three main pathways define the future of magnet materials. The first focuses on improving existing rare earth magnets. Techniques such as grain boundary diffusion, optimized sintering, and refined alloy microstructures can reduce dysprosium content while maintaining high-temperature performance. Such innovations ease pressure on heavy rare earth supply and lower magnet costs without requiring dramatic changes to manufacturing infrastructure.

The second pathway aims to develop rare-earth-free magnets. Materials such as manganese-aluminum-carbon, iron-nitrogen ($Fe_{16}N_2$), and cobalt-iron alloys show promising magnetic properties. Some offer energy densities approaching commercial rare earth magnets, though challenges remain in stability, manufacturability, and performance under thermal stress. Research institutes and industry partners worldwide are pushing these materials forward, with the hope that breakthroughs in processing or microstructure control may unlock their full potential. If successful, rare-earth-free magnets could fundamentally reshape the motor industry and reduce dependence on China's magnet supply chain.

A third pathway explores entirely new classes of magnetic materials, including magnetic metamaterials and atomically engineered structures. These advanced materials use lattice geometry and nanoscale design to achieve magnetic performance through structure, not composition. Although still largely experimental, they represent a radical shift in how magnetic properties could be engineered in the future.

Batteries are undergoing an even more dramatic transformation. For decades, lithium-ion has been the undisputed leader in energy storage for electronics, EVs, and increasingly the grid. Yet rising demand for lithium, nickel, cobalt, and graphite—combined with geographic concentration and environmental concerns—is accelerating the search for alternative chemistries.

Lithium-ion itself is evolving. Nickel-rich chemistries offer higher energy density, while cobalt-free or low-cobalt variants address supply risks. Lithium iron phosphate (LFP), with no nickel or cobalt, is scaling rapidly and is poised to dominate mass-market EVs and stationary storage due to its cost, safety, and cycle life advantages. Lithium manganese iron phosphate (LMFP) is emerging as an intermediate option that improves energy density while preserving LFP's safety and affordability.

Beyond lithium-ion, sodium-ion batteries are entering commercial deployment. Sodium is abundant, geographically distributed, and inexpensive. While sodium-ion batteries offer lower energy density, they are ideal for grid storage, two-wheelers, low-cost vehicles, and industrial applications. Their rapid commercialization could significantly diversify global battery supply chains and relieve pressure on lithium markets.

Solid-state batteries represent one of the most anticipated future technologies. By replacing liquid electrolytes with solid materials, they promise higher energy density, improved safety, and longer lifespans. However, solid-state batteries require new materials—ceramic electrolytes, lithium metal anodes, and advanced cathode structures—that introduce their own supply challenges. Scaling solid-state technology will require breakthroughs in materials processing, interface engineering, and manufacturability. Yet if these challenges are overcome, solid-state batteries could redefine the energy landscape.

Battery innovation is also branching into exotic chemistries. Lithium-sulfur batteries offer theoretical energy densities far above current technologies while using sulfur—a plentiful material. Their main challenge is cycle life. Zinc-air, aluminum-air, and other metal-air chemistries promise extremely low-cost, high-energy storage but require major advances in catalysts and durability. Vanadium flow batteries, though limited by vanadium supply, offer long-duration storage with excellent recyclability, making them attractive for grid applications. These technologies reflect a broader truth: the battery future will not be defined by a single chemistry but by a diverse ecosystem optimized for different applications.

On the anode side, the shift from graphite to silicon anodes is one of the most significant pending material transitions. Silicon can store far more lithium per unit mass than graphite, enabling higher energy density. Many next-generation EV batteries will likely incorporate silicon-dominant anodes, though challenges related to expansion, degradation, and binder chemistry remain. The transition to silicon anodes will also shift material demand patterns, reducing pressure on graphite supply chains and increasing interest in high-purity silicon feedstocks.

Recycling will become an essential part of the battery future. As early EV fleets reach end-of-life, recycled nickel, cobalt, lithium, and manganese will become major inputs for new batteries. Advanced direct recycling methods, which preserve cathode structures, could significantly reduce energy consumption and material loss. Over time, a circular battery ecosystem may stabilize material markets and reduce geopolitical risk.

In semiconductors, the future of materials will be defined by the limits of silicon and the rise of compound and two-dimensional semiconductors. Silicon has been the foundation of computing for more than half a century, yet it is approaching its physical limits. To continue scaling performance, the semiconductor industry is turning to materials with superior electrical, thermal, and optical properties.

Wide-bandgap semiconductors such as gallium nitride (GaN) and silicon carbide (SiC) will dominate power electronics for EVs, renewable energy systems, 5G infrastructure, and next-generation industrial equipment. Their efficiency advantages reduce energy losses and enable compact, high-performance systems. Demand for high-purity SiC wafers and GaN substrates will therefore grow exponentially. These materials require specialized production processes and depend on supply chains that are far less developed—and more fragile—than those of silicon.

Compound semiconductors such as indium phosphide (InP) and gallium arsenide (GaAs) will underpin the future of optical communication, quantum technologies, and high-frequency electronics. As global data consumption grows, telecom networks will require more indium- and gallium-based chips to power high-speed, high-bandwidth connections. Quantum computing systems will rely on unique materials such as ultra-pure niobium, sapphire substrates, isotopically purified silicon, ytterbium ions, or rare-earth-doped crystals—each requiring supply chains far removed from those of traditional semiconductors.

Two-dimensional materials represent another frontier. Graphene, molybdenum disulfide (MoS_2), and related layered compounds could enable ultrathin, flexible, and high-speed devices. While many remain in the research phase, they hold potential for sensors, transistors, optoelectronics, and quantum devices. If successfully commercialized, 2D materials could introduce entirely new sourcing and processing challenges, as well as new opportunities for material independence.

Photonic and neuromorphic computing may reshape future material demand even further. Photonic chips, which use photons instead of electrons to process information, require exotic materials such as silicon nitride, lithium niobate, and indium phosphide. Neuromorphic devices, designed to mimic human neural networks, require memristive materials, phase-change compounds, and novel oxide structures. These emerging architectures illustrate a future where computing performance depends heavily on mastering a wider array of advanced materials.

Across magnets, batteries, and semiconductors, a clear theme emerges: material systems are becoming more complex, not less. Innovation is expanding the periodic table of technology, not contracting it. Each new advancement introduces new dependencies, new supply chains, and new vulnerabilities. The challenge for the future is not to remove material complexity but to manage it responsibly—through diversified supply, advanced recycling, substitution, and strategic investment in refining and processing capacity.

The next stage of global technological development will depend on how effectively nations balance innovation with material security. Those who master the materials of the future—whether rare earths, advanced semiconductors, next-generation battery compounds, or novel magnetic alloys—will shape the trajectory of the twenty-first century.

PART IV – GEOPOLITICS, POWER, AND CONFLICT

CHAPTER 27 – RESOURCE NATIONALISM AND STATE STRATEGY

Resource nationalism is not a new phenomenon, but in the era of critical minerals it has taken on new intensity, new forms, and new geopolitical implications. As the world electrifies, digitizes, and militarizes its economies, states have begun to view key materials—lithium, cobalt, nickel, rare earths, copper, graphite, gallium, germanium, and many others—not as commodities but as instruments of national power. This shift is reshaping global politics, investment flows, industrial strategy, and diplomatic relations. Resource nationalism now exists not at the fringes of economic policy but at the center of statecraft. Understanding its drivers and implications is essential to understanding the geopolitics of the 2020s and beyond.

At its core, resource nationalism reflects a state's desire to capture greater control over the extraction, refining, and export of its natural resources. Historically associated with oil and gas, it now extends into the mineral systems underlying renewable energy, advanced manufacturing, and national defense. The logic is straightforward: if the world needs what you have, and supply chains are narrow and fragile, then controlling those resources becomes a lever of economic development, political influence, and strategic security.

Several factors are fueling the resurgence of resource nationalism. First is the geopolitical value of critical minerals. As energy systems shift away from fossil fuels, states recognize that minerals—not oil—will anchor future power structures. Nations endowed with abundant deposits seek to maximize domestic capture of the value chain, often by restricting exports of raw ore and encouraging local processing and manufacturing. Indonesia's nickel policies exemplify this strategy. By banning the export of raw nickel ore and requiring foreign companies to build smelters and refining plants domestically, Indonesia has rapidly evolved into a global hub for nickel processing, attracting billions of dollars of investment and reshaping the battery materials market.

A second driver is industrial policy. Many states see the current period as a once-in-a-century opportunity to industrialize or reindustrialize by leveraging mineral wealth. Governments increasingly require that mining projects include downstream processing, local procurement, technology transfer, and domestic workforce development. These policies aim to transform resource extraction into broad-based economic growth rather than allowing foreign companies to extract value without building long-term domestic capacity.

A third driver is fiscal pressure. As global economic volatility rises, many governments view mineral revenues as essential for domestic budgets. Countries may increase royalties, renegotiate contracts, or impose export taxes to capture more revenue from high-demand minerals. These measures can generate short-term gains but may deter foreign investment or slow project development if perceived as unpredictable.

A fourth driver is strategic security. Nations worry about losing control over materials essential to energy systems, digital infrastructure, or defense technology. As great-power competition intensifies, securing access to critical minerals becomes a national priority. Resource nationalism becomes a defense mechanism: a way to ensure that foreign actors cannot dominate strategic sectors or use resource dependencies as leverage.

Resource nationalism manifests in a variety of ways. Export controls are among the most visible tools. These may restrict raw material exports, impose tariffs on semi-processed goods, or require special licenses for strategic minerals. Indonesia's nickel ban, China's export controls on gallium and germanium, and Zimbabwe's restrictions on lithium ore exports all reflect this trend. Each has triggered ripple effects across downstream industries, highlighting how resource nationalism disrupts global supply chains that rely on open flows of materials.

Licensing rules and national ownership requirements are another tool. Some states require that foreign companies partner with domestic firms, transfer technology, or cede partial ownership of projects. Others mandate that state-owned enterprises participate in extraction or refining. This approach aims to ensure domestic control over strategic assets and prevent foreign monopolization of critical industries.

Local value-add mandates are increasingly common. Governments may require that minerals be processed domestically before export, pushing companies to invest in smelters, refineries, or fabrication plants. While these policies can stimulate industrial development, they require significant investment in infrastructure, energy, and human capital. Without these foundations, value-add mandates can backfire, stranding resources or reducing competitiveness.

Nationalization or renegotiation of mining contracts represents a more aggressive form of resource nationalism. While full nationalization is rare today, partial nationalization or renegotiation of terms is common. Countries such as Chile, Mexico, Bolivia, and Mongolia have reassessed mining laws to increase state control or expand state participation in mining ventures. The trend reflects a global rebalancing of power between resource-rich nations and corporations.

Environmental and social pressures also feed into resource nationalism. Communities affected by mining increasingly demand greater benefits, stronger protections, and more equitable distribution of profits. Governments, responding to domestic political pressure, may tighten environmental regulations, impose stricter permitting requirements, or halt projects entirely. These measures can protect ecosystems and local populations but may also slow global mineral supply growth.

Resource nationalism is particularly visible in Latin America, Africa, and parts of Asia—regions with vast mineral wealth and histories of extraction without significant domestic benefit. In Latin America, the so-called "lithium triangle" nations—Chile, Argentina, and Bolivia—are asserting greater control over lithium resources. While their strategies differ, each seeks to capture more value by promoting domestic processing and battery manufacturing. Bolivia has long resisted foreign investment that does not include significant state participation, while Chile is moving toward a public-private partnership model for strategic lithium operations.

In Africa, nations such as the Democratic Republic of Congo, Namibia, Tanzania, and Zimbabwe are tightening rules on cobalt, copper, graphite, and lithium. Some aim to build local processing industries; others seek to renegotiate terms with foreign mining firms. However, many African countries face governance challenges, limited infrastructure, and competition among foreign powers—all of which complicate the pursuit of resource sovereignty.

Southeast Asian states offer a contrasting picture. Indonesia's aggressive industrial policy has made it a dominant player in the nickel supply chain. The Philippines is considering similar strategies. Vietnam is positioning itself as a rare earth processing hub to attract global investment. These nations see resource nationalism not as isolationism but as a path to industrial integration.

China employs a different model of resource nationalism—less about restricting its own resources, more about securing overseas supply. Through foreign investment, long-term offtake agreements, and strategic financing, China has built a dominant global presence across cobalt, lithium, nickel, manganese, and rare earth supply chains. China's control does not stem solely from domestic resources but from outward-reaching state-backed strategy. Export controls, such as those on rare earths in the past and gallium/germanium more recently, demonstrate China's willingness to use its processing dominance as a geopolitical tool.

The resurgence of resource nationalism carries global implications. It reshapes supply chains by slowing or redirecting material flows. It increases investment risk for multinational corporations. It alters global pricing dynamics by creating bottlenecks or reducing supply elasticity. It intensifies geopolitical competition as major powers court resource-rich nations through diplomacy, aid, investment, and security guarantees.

For consuming nations—especially the United States, Europe, Japan, and South Korea—resource nationalism represents both a challenge and an opportunity. It complicates efforts to secure raw materials for clean energy and advanced manufacturing, but it also encourages the development of diversified, resilient supply chains. It pushes consuming nations toward alliances with resource-rich regions and promotes investment in recycling, substitution, and domestic production. It also highlights the need for a new type of international diplomacy centered on minerals, not just oil.

CHAPTER 28 – WEAPONIZATION OF SUPPLY CHAINS

The geopolitics of the twenty-first century is increasingly defined not by armies massing at borders or fleets confronting one another at sea, but by the invisible architecture of supply chains. As critical minerals become central to energy, technology, and national security, states are discovering that control over these supply chains can be used as leverage—sometimes subtle, sometimes overt, sometimes devastatingly effective. The weaponization of supply chains is not a hypothetical scenario; it is already happening, reshaping global markets, alliances, and strategic calculations. In this emerging landscape, access to materials becomes both a shield and a sword.

Weaponization in this context refers to the deliberate use of supply chain chokepoints, export controls, licensing restrictions, embargoes, or regulatory pressure to influence the behavior of other nations. Unlike traditional warfare, supply chain weaponization unfolds across financial systems, trade flows, logistics routes, and industrial dependencies. Its power lies in its ability to inflict economic pain, disrupt technological industries, or limit military capabilities without a single shot fired.

The roots of this trend extend from the simple fact that many critical minerals have extraordinarily concentrated supply chains. Rare earth separation, graphite purification, gallium refining, germanium production, lithium chemical conversion, nickel processing—so many of these steps depend on a handful of facilities in a handful of countries. When supply is concentrated, influence is automatic; when influence is deliberate, it becomes a weapon.

The 2010 rare earth crisis remains one of the earliest and clearest examples of supply chain weaponization. When China restricted rare earth exports to Japan amid diplomatic tensions, prices soared globally, industries panicked, and policymakers suddenly understood the strategic nature of mineral supply chains. Although the official justification was environmental enforcement, the geopolitical message was unmistakable: control over critical materials can be used as leverage. This episode reshaped global awareness and triggered policy responses across the United States, Europe, and Asia.

In the years that followed, supply chain weaponization spread to other materials. China's 2023 export licensing rules on gallium and germanium—minor metals essential for semiconductors, infrared optics, photovoltaics, and 5G—sent shockwaves through global industries. These controls were not embargoes but regulatory chokepoints enabling China to scrutinize, slow, or halt exports at will. Given China's overwhelming dominance in refining these materials, even slight interruptions ricocheted through semiconductor supply chains. The message was once again clear: control over niche materials grants disproportionate geopolitical power.

China is not the only actor capable of weaponizing supply chains. The United States uses export controls and technology restrictions to curb China's access to advanced semiconductors, chipmaking equipment, and certain defense-related materials. While these measures are framed in terms of national security, they function as strategic constraints—limiting China's technological advancement by targeting the materials, tools, and software it needs. Through alliances and coordinated restrictions, the United States has effectively internationalized its export control regime across Europe, Japan, and South Korea, demonstrating how advanced democracies are weaponizing supply chains in a defensive manner.

Russia, too, illustrates how materials become geopolitical tools. Following sanctions imposed after its invasion of Ukraine, Russia restricted exports of noble gases critical for semiconductor manufacturing. It also leveraged its dominant position in palladium and nickel to influence global markets. The resulting price spikes signaled that even countries traditionally considered commodity exporters can weaponize specific supply links when geopolitical tensions escalate.

Weaponization also takes softer forms through infrastructure control. When a country dominates key logistics corridors or secures ownership stakes in foreign ports, refineries, or mines, it gains the ability to apply pressure indirectly. China's global investments in ports, railways, and mineral assets give it visibility and influence over material flows. Control over logistics infrastructure can be weaponized through bureaucratic slowdown, selective inspection, or contractual manipulation—tools far less visible but often more effective than outright bans.

Private companies, although not states, play a role in supply chain weaponization as well. When a dominant supplier refuses to serve certain markets or prioritizes domestic demand, the effect mirrors geopolitical leverage. Large semiconductor equipment manufacturers, battery material suppliers, and rare earth magnet producers all have the ability to shape global access, sometimes at the behest of their governments, sometimes out of commercial necessity.

Weaponization can also occur through regulatory asymmetry. A government may impose stringent environmental or safety rules selectively to restrict exports or slow foreign access to processed minerals. These measures may be framed as environmental stewardship but can serve strategic goals. For example, tightening regulations on rare earth production allows a state to exert pressure on downstream industries without explicitly invoking geopolitics.

The consequences of supply chain weaponization are profound. For one, it destabilizes global markets. When a critical mineral becomes subject to geopolitical maneuvering, prices spike, long-term contracts rupture, and companies scramble to diversify. This volatility undermines industrial planning, increases manufacturing costs, and slows technological deployment. For sectors such as renewables, EVs, and semiconductors, even temporary disruptions can cascade into multi-year delays.

Weaponization also reshapes alliances. Countries seek partnerships not only based on ideology or geography but on mutual material dependence. New forms of "minerals diplomacy" emerge as states negotiate supply assurances, co-invest in mining projects, or develop shared refining capacity. At the same time, supply chain coercion can fracture alliances. Nations caught between competing powers may face pressure to align with one or the other based on access to key materials.

Another consequence is the acceleration of nearshoring, friendshoring, and reshoring. Companies and governments increasingly view long, geographically diverse supply chains as security liabilities. Instead, they seek supply within politically aligned nations or develop domestic capacity. While these strategies enhance resilience, they raise costs and risk creating fragmented global markets.

Weaponization also encourages stockpiling. Nations build strategic reserves of critical minerals not only to hedge against supply disruptions but to insulate themselves from foreign coercion. Stockpiles act as buffers in crises but require careful management, financing, and forecasting. If poorly coordinated, they can distort markets or trigger competitive hoarding.

The spread of supply chain weaponization raises critical questions about the future of globalization. The assumption that trade promotes interdependence—and therefore stability—is being tested. Minerals that were once quiet commodities are now strategic assets entwined in geopolitical competition. Trust, transparency, and openness—cornerstones of global trade—are increasingly undermined by mutual suspicion.

Yet weaponization is not an unstoppable trend. Many nations recognize that extreme coercion ultimately undermines their own credibility and harms global stability. Some are promoting cooperative frameworks for mineral governance, encouraging joint ventures, shared standards, and multi-country investment initiatives. These efforts seek to channel resource competition toward stable, rules-based systems rather than zero-sum conflict.

CHAPTER 29 – THE CHINA–U.S. CONTEST FOR MATERIAL SOVEREIGNTY

The struggle between China and the United States over critical minerals is shaping the architecture of global power in the twenty-first century. Unlike previous eras in which competition centered on oil, nuclear capability, or military presence, today's rivalry revolves around materials that are geographically concentrated, technologically indispensable, and deeply embedded in every advanced economic sector. Rare earths, lithium, cobalt, graphite, gallium, germanium, nickel, copper, and dozens of specialized metals now form the backbone of energy systems, artificial intelligence, advanced manufacturing, and national defense. Control over these materials is rapidly becoming synonymous with geopolitical influence. The contest for material sovereignty—defined as a state's ability to secure, control, and utilize the resources essential to its technological and military ambitions—now sits at the center of U.S.–China competition.

China's strategy for achieving material sovereignty began decades earlier than that of the United States. While Western economies shifted manufacturing overseas, China built a comprehensive industrial ecosystem spanning mining, refining, processing, and advanced manufacturing. It invested heavily in mineral-rich countries, supported state-owned enterprises across Africa and Latin America, and pursued long-term offtake agreements for everything from cobalt and lithium to copper and rare earth concentrates. Simultaneously, China poured capital into domestic refining capacity—the most strategically important and technically complex segment of the supply chain. Today, China processes the majority of global rare earth oxides, produces most of the world's graphite anodes, refines the majority of its lithium chemicals, dominates cobalt and nickel refining, and controls key intermediate materials used in magnets, batteries, solar panels, and electronics.

The United States, by contrast, entered the twenty-first century with an assumption of supply chain security and a belief in market-driven diversification. It allowed domestic mining to contract, ceded refining to lower-cost producers, and relied on global trade to provide the minerals needed for its digital and defense industries. Both government and industry underestimated how quickly supply could concentrate in a single nation—and how that concentration could be weaponized. Only in the past decade, and especially since the mid-2010s, has the United States recognized the strategic implications of this dependence.

The competing visions of material sovereignty between China and the United States reflect fundamentally different philosophies. China views minerals as strategic assets essential for national rejuvenation, technological dominance, and military capability. Its industrial policies, five-year plans, outbound investment strategies, and export controls are coordinated to secure long-term leverage. The United States, meanwhile, frames material sovereignty in terms of resilience, open markets, and national security. Its approach relies more heavily on alliances, private-sector incentives, and efforts to rebuild domestic capacity.

The rivalry is most visible in rare earths—perhaps the most iconic battleground of material sovereignty. China controls the vast majority of global rare earth separation and magnet production, giving it leverage over key components used in electric vehicle motors, wind turbines, missiles, radars, and precision-guided weapons. While the United States has reopened mining operations and initiated magnet manufacturing projects, it remains dependent on China for separation and high-purity oxides. Washington's strategic goal is to develop a fully domestic—or at least allied—rare earth value chain, reducing its vulnerability to Chinese export controls or geopolitical tensions.

Batteries form another core front in the contest. China dominates lithium refining, cobalt and nickel chemicals, anode and cathode production, and battery cell manufacturing. It has established a global lead in the gigafactory race, enabling unparalleled economies of scale. The United States responded with massive domestic incentives—via the Inflation Reduction Act and bipartisan industrial policies—to localize battery production and reduce reliance on Chinese materials. But building domestic gigafactories is only a partial solution; without secure access to processed lithium, nickel, cobalt, and graphite, battery sovereignty remains incomplete.

Graphite is perhaps the least discussed but most revealing material in the competition. China refines nearly all of the world's battery-grade graphite—anode material required in every lithium-ion battery. Even if the United States builds dozens of new battery factories, the absence of domestic graphite refining remains a major vulnerability. Both countries are now competing for alternative graphite sources in Africa, North America, and emerging synthetic graphite technologies.

The semiconductor supply chain, although dominated by Taiwan, South Korea, Japan, and the United States at the manufacturing level, still relies heavily on China for gallium, germanium, graphite, rare earth dopants, and dozens of upstream materials used in photolithography, wafer preparation, and packaging. China's export restrictions on gallium and germanium signaled that it would not hesitate to leverage control over these critical inputs when faced with U.S. technology sanctions. The U.S. response—extending semiconductor export controls, restricting access to advanced lithography tools, and tightening investment oversight—reflects the intensifying technological dimension of material sovereignty.

Both nations also compete geopolitically for influence over mineral-rich regions. In Africa, China has secured key mining positions in cobalt, copper, and manganese, while the United States attempts to build alternative partnerships through diplomatic engagement and "Miners for the Future" initiatives. In Latin America, China invests aggressively in lithium extraction and processing, while the United States seeks to counterbalance through trade agreements, developmental aid, and private-sector investment. Southeast Asia has become a battlefield for nickel, tin, and rare earth processing, with Indonesia and Vietnam playing both sides to maximize economic gain.

This global competition creates fragmented spheres of influence. China pursues an integrated strategy: build overseas mines, develop dominant refining capacity, incentivize companies to use Chinese materials, and create global dependence through high-volume manufacturing. The United States focuses on alliance-building. It courts Australia, Canada, Europe, Japan, and South Korea to develop secure mineral corridors, joint refining projects, and shared stockpiles. Rather than seeking total material autonomy, Washington aims to build a networked system of allied mineral supply chains insulated from Chinese control.

Material sovereignty also intersects with industrial policy. China ties its mineral strategy to its broader ambition to lead in electric vehicles, renewable energy, robotics, and AI hardware. Its goal is to secure both upstream inputs and downstream markets. The United States, on the other hand, seeks to revive domestic manufacturing after decades of offshoring. Strategic subsidies aim to bring semiconductor fabrication, battery assembly, and advanced manufacturing back to U.S. soil. But success depends on whether the U.S. can also secure upstream materials, something it has not yet achieved at scale.

The emerging contest is not simply economic—it is ideological. China promotes a state-centric, long-term strategy where the state directs capital, secures resources, and orchestrates supply chains. The United States advances a hybrid model combining market innovation, private-sector leadership, and government incentives. As these systems collide, the global market for critical minerals becomes a battleground for influence. Countries must choose whether to align with Chinese industrial ecosystems, U.S.-led alliances, or a nonaligned approach that extracts benefits from both sides.

Yet the contest for material sovereignty is not a binary confrontation. It evolves across shades of economic interdependence. China still relies on U.S., Japanese, and European technologies for certain manufacturing tools. The U.S. still relies on Chinese materials for many components. Decoupling, even partial, is difficult and costly. The most likely future is one of selective disentanglement: each nation securing independence in areas deemed strategically vital while maintaining global integration where substitution is difficult or prohibitively expensive.

The stakes of the China–U.S. contest for material sovereignty are enormous. The next decade will determine whether the world remains deeply interconnected or fractures into competing blocs with parallel supply chains. It will determine which nation leads the industries of the future, controls the technologies that define power, and shapes the rules of global trade. As the next chapter will show, this bilateral competition is only one part of a broader global scramble—one that extends across Africa, Latin America, and emerging mineral frontiers where new alliances, rivalries, and opportunities are rapidly unfolding.

CHAPTER 30 – AFRICA, LATIN AMERICA, AND THE NEW GLOBAL SCRAMBLE

A new global scramble for critical minerals is unfolding across Africa and Latin America—regions that hold some of the world's richest deposits of lithium, cobalt, copper, manganese, rare earths, graphite, nickel, and other essential resources. These regions were once peripheral to global industrial power, often serving as raw material suppliers to distant manufacturing centers. Today, however, they sit at the crossroads of the energy transition, industrial transformation, and geopolitical rivalry. China, the United States, the European Union, India, Russia, Gulf states, Japan, and South Korea are all competing—sometimes directly, sometimes subtly—for influence over the mineral systems that will underpin the future of global power.

Africa and Latin America together represent a unique combination of abundant geology, structural vulnerabilities, and rising political agency. They are no longer passive actors in the global resource economy; they are increasingly shaping agendas, renegotiating contracts, imposing new rules, and asserting sovereignty over their resources. In both regions, the global scramble is producing a complex web of partnerships, rivalries, and strategic calculations.

Africa is arguably the center of gravity for the new scramble. The continent hosts vast mineral wealth: cobalt in the Democratic Republic of Congo (DRC), platinum group metals in South Africa and Zimbabwe, manganese in South Africa and Gabon, graphite in Mozambique and Tanzania, rare earth deposits in Burundi, Madagascar, Angola, and South Africa, copper in Zambia and the DRC, and nickel in countries like Botswana and Madagascar. These minerals are essential for batteries, electric motors, solar panels, wind turbines, fuel cells, semiconductors, and advanced manufacturing.

The DRC remains the most strategically important African country for critical minerals due to its extraordinary cobalt reserves and significant copper output. More than half of the world's cobalt comes from the DRC, making it an indispensable part of the global battery ecosystem. China's dominance in DRC mining—through ownership stakes, offtake agreements, infrastructure-for-resource deals, and large-scale investment—illustrates how Beijing moved early and decisively. Chinese companies process much of the DRC's cobalt and copper, transporting concentrated feedstocks to China's vast refining complexes. This system locks African extraction into China's industrial frameworks.

The growing global competition for cobalt, copper, and associated minerals is pushing other nations to increase engagement in the region. The United States has launched diplomatic initiatives, supply chain partnerships, and development financing programs to counterbalance China's influence. Europe is mobilizing through the Global Gateway initiative, aiming to co-invest in mining, refining, and infrastructure. Japan and South Korea are partnering on critical mineral agreements, while Gulf countries—particularly the UAE and Saudi Arabia—have begun investing heavily in African mining and processing as part of their industrial diversification strategies.

Africa's governments are responding with new forms of resource nationalism and strategic assertiveness. Several countries have imposed export restrictions on unprocessed minerals, seeking to capture more value within their borders. Namibia banned unprocessed lithium and rare earth exports; Zimbabwe enacted similar measures for lithium; the DRC has revisited its mining code to increase state revenues; and Tanzania is pushing for local processing of graphite. These policies reflect a desire to avoid the historical pattern of material extraction with limited domestic benefit. However, they also create uncertainty for investors and require significant infrastructure, energy capacity, and technical expertise to succeed.

South Africa continues to be a major player in platinum group metals and manganese, but chronic electricity shortages, governance challenges, and infrastructure weaknesses hamper its ability to capitalize on rising global demand. New players such as Rwanda, Botswana, and Angola are emerging as mineral hubs, offering more stable governance and investor-friendly regulatory environments. The competition between African nations to attract investment creates a dynamic landscape, one where geopolitical players must navigate differing priorities, regulatory frameworks, and strategic visions.

Latin America represents the other major front in the global scramble. The region's lithium triangle—Argentina, Bolivia, and Chile—contains more than half of the world's known lithium resources. Chile and Argentina are among the largest lithium producers, while Bolivia possesses vast but technically challenging brines that remain underdeveloped. Lithium is not the region's only asset: Peru and Chile are top global producers of copper; Brazil is rich in nickel, niobium, bauxite, and rare earths; Mexico holds significant lithium deposits; and Colombia is exploring emerging copper and rare earth opportunities.

China's presence in Latin America mirrors its approach in Africa: early, strategic, and multidimensional. Chinese companies have invested heavily in lithium projects in Argentina and Chile, copper mines in Peru, and infrastructure that supports mining logistics. China has also extended financial support and offtake agreements to ensure stable access to regional mineral flows. Latin America's lithium sector, in particular, has become a focal point of U.S.–China competition. China aims to secure supply for its battery manufacturing system, while the United States seeks to reduce dependence on Chinese materials.

Latin American governments are increasingly asserting control over their minerals. Chile has outlined a national lithium strategy that increases state participation in strategic assets, though not eliminating private involvement. Argentina remains more open to foreign investment but is tightening regulatory oversight. Bolivia has pursued a long-standing state-led model with limited foreign involvement, though recent partnerships signal a possible shift toward joint ventures. Mexico has nationalized its lithium sector and created a state-owned enterprise to manage it, though the practical implementation remains uncertain.

Beyond lithium, Latin America's copper sector is drawing intense global interest. Copper is essential for electrification, grid expansion, EVs, and renewable energy. Chile, Peru, and Mexico host some of the largest copper mines in the world, but political instability, water scarcity, and community resistance have slowed development. These factors introduce geopolitical and supply chain risks at a time when global copper demand is expected to surge.

Brazil, with its diversified mineral profile, is becoming a key player in the global scramble. It is the world's dominant producer of niobium—a metal essential for strengthening steel—and holds significant deposits of nickel, graphite, rare earths, and bauxite. Brazil's stable industrial base and deep mining expertise position it as a future powerhouse in critical minerals, though environmental concerns—particularly in the Amazon—present political and regulatory challenges.

Both Africa and Latin America are leveraging the geopolitical rivalry between China and the United States to their advantage. Governments are negotiating higher royalties, more stringent environmental standards, domestic value-add requirements, and local employment conditions. They are also using strategic ambiguity—engaging multiple partners simultaneously—to maximize economic benefits. These regions understand that critical minerals give them negotiating power unprecedented in previous eras of resource exploitation.

Yet the new scramble is not solely defined by major-power competition. India is expanding partnerships in Africa and Latin America to secure lithium, cobalt, nickel, and copper for its growing manufacturing sectors. European states are seeking long-term agreements to support their clean energy industries. Gulf nations, seeking economic diversification, are investing in mining, refining, and battery supply chain assets across both continents. South Korea and Japan are pursuing joint ventures, offtake agreements, and technology partnerships to ensure stable supply for their automotive and electronics sectors.

This multipolar competition creates opportunities for resource-rich nations but also carries risks. Fragmentation of global supply chains may increase costs and reduce efficiency. Overly aggressive resource nationalism could deter investors and delay critical projects. Weak governance or political instability could derail large-scale investments. Environmental and social conflicts could escalate as mining expands into sensitive areas, from the Congo Basin to the Lithium Triangle.

Still, Africa and Latin America are no longer spectators in global resource politics—they are active participants shaping the future of material flows, industrial ecosystems, and geopolitical alignments. Their decisions will influence the cost and speed of the energy transition, the stability of global supply chains, and the balance of power among major nations.

Alliances and partnerships are becoming instruments for managing material risk as much as traditional security concerns. Through supply agreements, shared investments, and emerging forms of minerals diplomacy, states seek to reduce exposure while shaping access to strategic resources. The contest over critical materials is increasingly negotiated through coalitions rather than markets alone.

CHAPTER 31 – ALLIANCES, PARTNERSHIPS, AND "MINERALS DIPLOMACY"

The global competition for critical minerals is no longer fought solely through markets or bilateral trade agreements. It has evolved into a complex system of alliances, partnerships, and diplomatic strategies, collectively forming what can be called minerals diplomacy. Nations are reconfiguring their geopolitical relationships around the need to secure access to the materials that power electric vehicles, renewable energy systems, semiconductors, defense platforms, and advanced manufacturing. In this environment, cooperation is as strategically important as competition, and states increasingly rely on multilateral frameworks, shared investments, and coordinated policy initiatives to shape the contours of the emerging mineral order.

Minerals diplomacy begins with the recognition that no country is self-sufficient. Even nations with significant geological endowments lack refining capacity, processing expertise, or manufacturing ecosystems. Others possess industrial might but limited mineral reserves. This asymmetry pushes governments to build alliances that align resources, technology, capital, and geopolitical interests. The United States, the European Union, Japan, South Korea, Australia, Canada, and India are all pursuing partnerships explicitly designed to counterbalance China's dominance across mineral supply chains.

Among these, the United States-led Minerals Security Partnership (MSP) is the most prominent. Established to coordinate investment and policy among major democracies, the MSP includes allies such as Australia, Canada, Japan, South Korea, the United Kingdom, and the European Union. The partnership's goal is to accelerate strategic mining and refining projects, support sustainable development practices, and ensure supply chain transparency. MSP members pool financial resources, align regulatory frameworks, and coordinate overseas investments to provide alternatives to Chinese-backed mining ventures. Though still in early stages, MSP has already signaled a shift toward collective action—turning mineral security into an alliance-wide priority rather than a purely national one.

The United States has also deepened bilateral mineral agreements with key partners. Its agreements with Australia and Canada focus on integrating North American and Australian resources into shared supply chains for batteries, defense technologies, and critical manufacturing. These deals combine financial incentives with regulatory harmonization to encourage cross-border cooperation. The U.S.–Japan partnership focuses heavily on semiconductor materials and rare earths, while the U.S.–South Korea alliance emphasizes battery minerals and industrial collaboration. These partnerships reflect Washington's belief that no single nation—even one with vast technological and financial capabilities—can compete against China's mineral strategy alone.

Europe has taken a different but complementary approach. Through its Critical Raw

Materials Act, the European Union seeks to build a strategic framework for sourcing, processing, and recycling critical minerals from both domestic and international partners. The EU's Global Gateway initiative provides funding for mining infrastructure, power systems, and transportation corridors across Africa, Latin America, and the Indo-Pacific. These investments are designed not only to secure materials but also to advance European standards of sustainability, governance, and transparency. The EU's approach blends economic diplomacy with normative influence, aiming to counter China's dominant presence by promoting responsible mining and ethical supply chains.

Japan and South Korea, both heavily dependent on imported minerals, have embraced minerals diplomacy with exceptional urgency. Japan's advanced industries require stable supplies of rare earths, gallium, germanium, and battery materials. After suffering disruptions during the 2010 rare earth incident, Japan diversified aggressively—developing mining partnerships in Vietnam, India, Malaysia, and Australia, investing in recycling technologies, and supporting domestic magnet manufacturing. Japan's long-term strategy emphasizes resilience through diversification, technological innovation, and stockpiling.

South Korea, home to major battery and electronics manufacturers, has structured its minerals diplomacy around industrial logic. Korean companies often negotiate directly with governments and mining firms across Africa, Latin America, and Southeast Asia. The Korean government supports these corporate initiatives with diplomatic backing, insurance guarantees, and strategic financing. This public-private model allows South Korea to build robust, vertically integrated supply chains without relying exclusively on state-led initiatives.

Australia and Canada play unique roles as both mineral suppliers and geopolitical partners. Their abundant resources, stable political environments, and strategic alignment with the United States make them central pillars of Western minerals diplomacy. Australia's rare earth, lithium, and nickel sectors have become focal points of allied investment, while Canada's nickel, cobalt, graphite, and copper deposits are being integrated into North American supply chains. Both countries participate actively in MSP and other cooperative frameworks, acting as resource anchors within broader alliance structures.

India is an increasingly important actor in minerals diplomacy. With rapid industrial expansion and ambitious manufacturing goals, India needs vast quantities of critical materials. Its diplomatic strategy emphasizes diversification away from China, forming partnerships with Australia, Japan, the United States, and African states. India's agreements with countries such as Namibia, South Africa, and Zambia aim to secure access to lithium, cobalt, and rare earths. At the same time, India is positioning itself as a future processing hub, offering an alternative to China's refining dominance.

The Gulf states are emerging as unexpected but influential players in minerals diplomacy. Saudi Arabia, the UAE, and Qatar are investing in global mining assets to diversify their economies beyond fossil fuels. Their sovereign wealth funds have begun financing mining ventures in Africa, Latin America, and Central Asia. They also seek to develop domestic refining capacity and advanced materials industries. This involvement adds a new dimension to the geopolitics of minerals, as Gulf states begin to act not just as investors but as strategic brokers.

China, however, remains the most sophisticated practitioner of minerals diplomacy. Its strategy is characterized by long-term engagement, infrastructure-for-resource deals, state-backed financing, and integration of mining investments into broader economic initiatives such as the Belt and Road Initiative. In Africa, Latin America, and Southeast Asia, Chinese companies often offer rapid project execution, favorable financing terms, and integrated logistics networks. These incentives are attractive to many governments seeking development gains. China's diplomacy is therefore not merely commercial; it is embedded in an overarching geopolitical architecture designed to secure long-term supply and influence.

As alliances proliferate, competition inevitably intensifies. Minerals diplomacy often leads to overlapping claims, conflicting standards, and rival investment proposals. Governments in Africa and Latin America exploit this competition to negotiate better terms, raise royalties, push for local processing, or demand technology transfer. While this can yield benefits, it can also increase complexity and introduce geopolitical tension into domestic policy decisions.

Despite competitive pressures, minerals diplomacy also creates space for cooperation on sustainability, human rights, and environmental standards. Many alliances emphasize ethical supply chains, anti-corruption measures, and reductions in carbon intensity. These shared commitments can help elevate global standards, though implementation remains uneven.

Minerals diplomacy is reshaping the international order in three fundamental ways. First, it is redefining alliances. States form partnerships not only based on ideology or security but increasingly based on shared material needs. Second, it is redistributing power. Resource-rich nations now wield unprecedented leverage, while industrial powers must negotiate access to resources they once took for granted. Third, it is transforming development trajectories. Countries with mineral wealth are using this moment to pursue industrialization, strategic autonomy, and geopolitical relevance.

When diplomacy reaches its limits, material flows become instruments of coercion. Sanctions, countersanctions, and market leverage are increasingly applied to reshape access, influence, and compliance. In the domain of critical minerals, economic pressure now operates as a direct extension of geopolitical power.

CHAPTER 32 – SANCTIONS, COUNTERSANCTIONS, AND MARKET LEVERAGE

Sanctions have become one of the defining instruments of geopolitical power, and in the era of critical minerals they carry even greater strategic weight. Unlike traditional trade measures, mineral-related sanctions target the foundational materials that support entire industries—energy, defense, electronics, and advanced manufacturing. In this new landscape, sanctions and countersanctions do not merely restrict access; they destabilize supply chains, shift global investment patterns, and reshape long-term alliances. Critical minerals amplify the leverage states can exert, because disruptions in seemingly small markets—gallium, germanium, palladium, titanium, or rare earth magnets—can have outsized impacts on high-tech systems. As a result, sanctions have become both sharper and riskier tools of economic statecraft.

The war in Ukraine brought these dynamics into sharp focus. When Western states imposed sweeping sanctions on Russia in 2022, they faced a dilemma: how to punish a major geopolitical aggressor without crippling their own access to key materials. Russia is one of the world's largest producers of palladium, a metal essential for catalytic converters in gasoline vehicles; its Norilsk operations supply roughly a third of global supply . It is also a key supplier of titanium used in commercial aerospace—Boeing and Airbus had long relied on Russia's VSMPO-Avisma for high-performance titanium components used in aircraft structures . Because of this dependence, Western sanctions initially avoided palladium and titanium, illustrating the strategic limits of sanction policy. Governments sought to isolate Russia financially while ensuring that global industries would not collapse under their own punitive actions.

Russia, for its part, possessed tools of retaliation. It temporarily restricted exports of noble gases—neon, argon, and helium—in 2022, disrupting global semiconductor fabrication because neon is essential for lithography lasers . Before the war, Ukraine had supplied roughly half of the world's purified neon, relying on byproducts from Russian steel mills. The conflict severed these connections, forcing chipmakers to diversify their sourcing and highlighting how quickly a localized war can ripple through supply chains. Though Russia has not fully weaponized palladium or titanium exports—likely because it relies on the revenue—its selective curtailments sent a clear signal: mineral producers can wield supply as leverage.

China has been even more deliberate in deploying countersanctions in response to Western technology restrictions. In 2023, after the United States and Europe tightened export controls on advanced semiconductors and chipmaking tools, China introduced export licensing requirements for gallium and germanium—two metals essential for compound semiconductors, optical systems, and high-frequency electronics. China controls roughly 80 percent of global gallium output and about 70 percent of germanium production . The restrictions immediately sent shockwaves through the semiconductor sector, causing price spikes, supply shortages, and bureaucratic delays as companies struggled to obtain permits. In December 2024, China escalated further by banning gallium and germanium exports specifically to the United States, forcing American firms to reroute procurement through intermediaries and exposing the fragility of their supply chains . Even though China later suspended the ban during trade negotiations, the episode demonstrated the potency of mineral countersanctions as instruments of geopolitical messaging.

China's leverage extends beyond gallium and germanium. In October 2023, it imposed licensing requirements on certain graphite products used in electric vehicle anodes—again citing national security concerns. Because China processes nearly all of the world's battery-grade graphite, this move was widely interpreted as a strategic warning to the United States and Europe about their dependence on Chinese refining capacity . Such measures reveal China's position as a "resource gatekeeper," capable of expanding or contracting the flow of critical materials to shape global behavior. Its early investments in refining, processing, and downstream manufacturing—combined with looser environmental regulations—have allowed it to become the indispensable intermediary in many mineral supply chains .

Sanctions and countersanctions interact with market leverage, a more diffuse but equally powerful force. In highly concentrated markets, the mere threat of disruption can trigger hoarding, price spikes, and emergency stockpiling. Following Russia's invasion of Ukraine in early 2022, industries scrambled to accumulate palladium in fear that it might be cut off, causing temporary price volatility and supply distortions . When Chinese export controls on gallium and germanium were announced, companies sought to secure alternative sources—even though very few existed—further intensifying market strain. The interplay between sanctions and market psychology can magnify the impact of political decisions far beyond their immediate scope.

Sanctions also reveal how vulnerable consuming nations are to supply chain chokepoints. The 2010 rare earth episode made this clear: when China halted shipments to Japan during a diplomatic dispute, prices soared by up to tenfold and industries were forced to restructure their material usage, launching recycling programs and thrifting efforts to reduce dependence on Chinese supply . The incident, although brief, reshaped strategic thinking across Asia, Europe, and North America. It demonstrated that even a small market could create systemic disruption, and that alternative supply lines take years to develop.

The threat of sanctions can itself be a tool. When the United States considered sanctions on Chinese rare earth magnets in 2019, China reacted with caution—not because of direct economic damage, but because such measures would accelerate diversification efforts in the West . Sanction threats force adversaries to reveal their vulnerabilities, disrupt long-term planning, and alter investment flows. Companies may shift supply chains preemptively to hedge against risk, even if sanctions are never implemented.

Market leverage does not belong exclusively to producers. Consuming nations can coordinate responses that weaken the effectiveness of producer sanctions. The G7's 2023 launch of a "Coordination Platform on Critical Minerals" aims to monitor markets collectively, prevent manipulation, share information on stockpiles, and develop synchronized responses to coercive actions—mirroring the International Energy Agency's role in oil markets . This initiative reflects a growing recognition that coordinated consumer action can counterbalance producer leverage.

Still, the asymmetry between producer and consumer power remains stark. Many critical minerals are byproducts of other mining processes—gallium from bauxite, germanium from zinc, iridium from platinum-group mining—meaning production cannot easily be increased in response to sanctions or shortages . This structural rigidity enhances the leverage of dominant producers: if China restricts gallium exports, the rest of the world cannot quickly increase supply. Similarly, if Russia halted palladium exports, global auto manufacturing would face immediate disruption.

Emerging resource nationalism can amplify sanction-related vulnerabilities. Countries such as Indonesia, Zimbabwe, and Chile have imposed export bans or restrictions to promote domestic processing or increase government revenue . While these measures reflect legitimate development goals, they also create additional chokepoints in global supply chains. For consuming nations, distinguishing between economic nationalism and geopolitical coercion becomes increasingly difficult as both can have similar disruptive effects.

Sanctions can also incentivize strategic shifts that extend far beyond the targeted materials. Following Western sanctions on Russia, aircraft manufacturers diversified their titanium supply chains, turning to Japan and domestic producers. Semiconductor firms expanded neon production outside Ukraine and Russia. Battery manufacturers accelerated research on alternative chemistries to reduce reliance on cobalt or nickel. These responses illustrate how sanctions shape long-term technological trajectories, sometimes reducing demand for the very materials used as leverage.

As tensions rise, scenarios of more severe mineral weaponization become plausible. China could restrict rare earth magnets in response to a conflict over Taiwan, severely affecting U.S. and European defense production—an F-35 fighter jet contains hundreds of kilograms of rare earth materials, and no rapid substitution is possible . Russia could weaponize palladium or titanium exports if geopolitical conditions worsen. Major cobalt or nickel producers could impose selective embargoes during diplomatic disputes. Any such action would reverberate across global industries, accelerating the formation of parallel supply chains and deepening geopolitical divisions.

Sanctions and countersanctions in critical minerals therefore operate within a wider strategic landscape—one in which economic statecraft is deployed alongside alliances, resource nationalism, and technological competition. The next chapter will explore how these forces converge into a broader "Critical Minerals Cold War," where blocs form, supply chains bifurcate, and the struggle for material power becomes a central feature of twenty-first-century geopolitics.

CHAPTER 33 – THE CRITICAL MINERALS COLD WAR SCENARIO

The world is entering a phase in which competition over critical minerals increasingly resembles a new kind of Cold War. Unlike the bipolar military confrontation of the twentieth century, this emerging conflict is defined not by ideological blocs or nuclear standoffs, but by control of supply chains, technological dominance, industrial capacity, and access to the materials that underpin the global economy. Critical minerals—once obscure to all but geologists, metallurgists, and niche industries—have become strategic assets capable of shifting the balance of power between nations. The scenario that is unfolding is not hypothetical; it is already taking shape in the policies, alliances, sanctions, and strategic investments of today's major powers.

A Critical Minerals Cold War emerges from the convergence of five structural realities. First, the global energy transition requires vast amounts of minerals—lithium, nickel, cobalt, copper, rare earths, graphite, manganese, and many others—that are more geographically concentrated than fossil fuels. Second, refining and processing capacity is even more concentrated, with China commanding a dominant position across nearly every critical mineral value chain. Third, geopolitical tensions between the United States and China have escalated into a long-term contest over technological supremacy, industrial leadership, and national security. Fourth, supply chains are fragile, vulnerable to disruptions, and increasingly shaped by political decisions rather than pure market forces. And fifth, many mineral-rich nations are asserting sovereignty through resource nationalism, complicating global access and creating new chokepoints.

Together, these forces create the conditions for a Cold War–like dynamic: rival blocs competing for influence, resources, and technological leadership, with the potential for fragmentation of global markets and parallel supply chains.

One hallmark of the Critical Minerals Cold War is the emergence of competing industrial ecosystems. In one bloc, China has built a vertically integrated system encompassing mining, refining, component manufacturing, and finished products—from batteries to electric vehicles, from solar modules to permanent magnets. Its dominance stems from decades of coordinated policy, state-backed investment, long-term global partnerships, and relentless industrialization. China's advantage lies not only in its control of raw materials but in its capacity to process them into advanced components. Even when minerals are mined elsewhere, they often end up in Chinese refineries.

The United States, Japan, South Korea, Europe, Australia, Canada, and India form the emerging counter-bloc—aligned by shared concerns about overdependence on China and by mutual interests in developing alternative supply chains. This alliance is not formalized in the manner of Cold War military blocs, but it manifests through coordinated policies, joint investments, shared technologies, and diplomatic frameworks. These nations are working to rebuild domestic mining, expand refining capacity, develop alternative supply corridors, fund recycling infrastructure, and invest in next-generation materials.

The contest between these blocs is intensifying in several key arenas. The first is technology. Advanced batteries, high-efficiency motors, compound semiconductors, quantum computing, and renewable energy systems all require highly specialized materials. Whoever controls the supply of these materials, and the factories that process them, shapes the trajectory of technological development. This has led to a wave of industrial policies such as the U.S. Inflation Reduction Act, Japan's subsidy programs for magnet production, Europe's Critical Raw Materials Act, and India's Production-Linked Incentive schemes. These initiatives aim to localize production, secure supply, and reduce dependence on foreign processing.

The second arena is geography. Africa, Latin America, and Southeast Asia—regions with significant mineral wealth—have become battlegrounds in the competition for influence. China's long-established partnerships, infrastructure investments, and offtake agreements give it a head start, but the United States, Europe, Japan, and India are rapidly increasing their diplomatic and financial engagement. This creates a multipolar scramble reminiscent of earlier eras, but with new power dynamics and far higher stakes. The Critical Minerals Cold War plays out not only between great powers but within the Global South itself, as governments negotiate terms, exercise sovereignty, and leverage competing interests to advance national goals.

The third arena is economic coercion. Export controls, sanctions, and countersanctions are now regular features of mineral geopolitics. China's restrictions on gallium, germanium, and graphite showed how materials can be weaponized. The United States' semiconductor export controls demonstrated that access to technology can be used as leverage. Russia's manipulation of noble gas and PGM supplies highlighted how conflict affects mineral markets. These tools are not isolated incidents—they are part of a broader pattern in which economic pressure is used to influence geopolitical behavior. In a Critical Minerals Cold War, such tools may become more frequent, more targeted, and more disruptive.

The fourth arena is innovation. Each bloc is racing to develop technologies that reduce or eliminate reliance on vulnerable minerals. Rare-earth-free magnets, cobalt-free batteries, sodium-ion batteries, alternative anodes, wide-bandgap semiconductors, and advanced recycling systems all form part of this technological arms race. Innovation becomes both a shield—protecting domestic industries from supply disruption—and a weapon—undermining the strategic leverage of rivals. The nation or bloc that leads in materials science may fundamentally reshape the geopolitics of the future.

The fifth arena is ideology—though not ideological in the Cold War sense. Instead of capitalism versus communism, the divide concerns industrial philosophy: China's state-driven model of long-term planning, coordinated investment, and vertically integrated supply chains versus the more decentralized, market-driven system of the United States and its allies, supplemented by strategic incentives and partnerships. This ideological divergence shapes how each bloc approaches resource security, trade, and technological development. It also influences how resource-rich nations align themselves.

A defining feature of a Cold War scenario is the potential bifurcation of global supply chains. In such a world, China would maintain a dominant supply chain for its domestic market and for aligned nations, while the United States and its partners would build parallel systems. These systems would differ in extraction, refining, environmental standards, technology transfer rules, and pricing mechanisms. Trade between blocs would shrink, and certain materials or technologies might become largely confined within one sphere or the other. This fragmentation would raise costs, slow technological deployment, and reduce global efficiency—but it would also reduce vulnerability to coercion.

The risk is that bifurcation hardens into an irreversible divide. If the world shifts into fully separate supply chain blocs, cooperation becomes harder, trust evaporates, and every mineral-rich region becomes a site of strategic competition. In this extreme scenario, investment, trade, and diplomacy are all filtered through the lens of mineral security. Countries that attempt to remain neutral may find themselves pressured or incentivized to choose sides.

Yet the Critical Minerals Cold War scenario is not predetermined. There are countervailing forces pushing toward cooperation rather than fragmentation. Many mineral-rich nations prefer a multipolar environment where they can negotiate with all sides. Global industries, especially in automotive and electronics, rely on economies of scale that favor integrated markets. Environmental concerns and sustainability goals push for shared standards and collaborative approaches. And even the major powers recognize that some level of interdependence remains unavoidable.

Still, the structural drivers of rivalry are strong, and the momentum is toward greater strategic competition. As the world electrifies and digitizes, the struggle for material power will intensify. Future flashpoints may arise around rare earth export controls, lithium nationalization, cobalt taxation, sanctions on semiconductors, or conflicts in mineral-rich regions like the DRC or the Lithium Triangle. In a world shaped by material scarcity and geopolitical pressure, even small disruptions can escalate into broader crises.

Tensions over critical minerals have already moved from abstraction to confrontation. Recent cases show how supply disruptions, policy shocks, and strategic responses cascade through markets and diplomatic channels. These episodes offer an early signal of how material constraints are reshaping geopolitical conflict.

CHAPTER 34 – CASE STUDIES: FLASHPOINTS FROM 2010 TO 2025

Between 2010 and 2025, the world witnessed a series of mineral-related crises, confrontations, and geopolitical ruptures that revealed the fragility and strategic importance of critical materials. These flashpoints were not random disruptions. They were early signals of a global system shifting toward resource competition, industrial vulnerability, and emerging forms of coercive power. Taken together, they chart the evolution of mineral geopolitics from a niche concern to a defining force shaping international relations. This chapter examines the most significant of these events, showing how each contributed to the broader dynamics explored throughout this book.

The 2010 rare earth crisis remains the foundational flashpoint of the critical minerals era. During a maritime dispute between China and Japan, China reportedly halted shipments of rare earths bound for Japanese customers. Whether this was an official embargo or an informal tightening of customs procedures remains debated, but the outcome was unambiguous: global rare earth prices skyrocketed, supply chains panicked, and governments began reassessing their exposure to mineral dependencies. The crisis exposed a structural vulnerability—China's dominance in rare earth processing—and demonstrated that minerals could be deployed as instruments of geopolitical pressure. It also triggered long-term diversification efforts in Japan, the United States, Europe, and Australia, planting the seeds for today's broader mineral strategies.

From 2013 to 2015, China's internal consolidation of its rare earth sector became another flashpoint with global implications. Facing environmental degradation, illegal mining, and chaotic overproduction, Beijing restructured its industry under a handful of large state-owned enterprises. While these reforms were motivated partly by domestic concerns, they also strengthened China's grip on the global rare earth market. The consolidation enabled more disciplined production, tighter export oversight, and greater state coordination—tools that would later prove influential during geopolitical tensions. This restructuring showed how domestic industrial policy in one country could affect global access to essential materials.

In 2014 and 2015, the annexation of Crimea and subsequent sanctions on Russia revealed another front in mineral geopolitics. Russia is a major producer of palladium, nickel, aluminum, potash, and titanium. As Western nations imposed financial and trade sanctions, industries dependent on Russian minerals became increasingly anxious about supply continuity. Particularly concerning were palladium, used in automotive catalytic converters, and titanium, essential for aerospace manufacturing. Although Russia did not fully weaponize these materials during the conflict, the episode illustrated how geopolitical shocks could jeopardize mineral supply chains and how sanctions could create secondary effects beyond their intended targets.

A significant but often overlooked flashpoint emerged in 2015 with Myanmar's rising role in global rare earth supply. The discovery and expansion of ionic clay deposits—a form of easily processed rare earth ore—rapidly positioned Myanmar as a major supplier of heavy rare earths. However, extraction was frequently linked to illicit operations and cross-border smuggling networks. When Myanmar experienced political turmoil and China increased enforcement at the border, global heavy rare earth supply tightened abruptly. This demonstrated how instability in even one small supplier can affect high-value segments of mineral markets and how China's cross-border trade decisions can reverberate internationally.

The years 2016 to 2019 saw escalating trade tensions between China and the United States, culminating in the U.S. imposition of tariffs on Chinese goods and China's threat to restrict rare earth exports. Although the rare earth restrictions were never enacted, the mere threat was enough to cause global concern. China's state media hinted at using rare earths as leverage, drawing a direct parallel between mineral supply chains and national security. This rhetorical escalation signaled a shift: minerals were no longer being used only informally or opportunistically, but were now entering the discourse as explicit strategic tools.

In 2017 and 2018, the rise of Indonesia's resource nationalism produced a different type of flashpoint. In an effort to build domestic industrial capacity, Indonesia banned the export of unprocessed nickel ore and incentivized foreign companies to construct local smelters. This abrupt policy change shocked the global nickel market and caused price volatility, with ripple effects across stainless steel, battery materials, and industrial metals. Indonesia's move demonstrated how resource-rich states could reshape global supply chains through policy alone, using sovereign control over minerals to extract long-term economic advantage. The success of Indonesia's strategy inspired similar measures in other nations, accelerating a global trend toward resource nationalism.

The COVID-19 pandemic in 2020 further exposed mineral vulnerabilities. Disruptions in shipping, labor availability, border controls, and industrial activity affected everything from copper production in Chile and Peru to lithium output in Argentina, nickel refining in Southeast Asia, and rare earth separation in China. Lockdowns temporarily shut down major mining hubs, revealing just how tightly the world depended on uninterrupted material flows. The pandemic also strained semiconductor supply chains, prompting a global shortage that highlighted the importance of materials like gallium, germanium, neon, and high-purity silicon. It became evident that disruptions in one layer of the chain could cascade into crises for technology, automotive manufacturing, and defense.

In 2021, energy shortages in China created another unexpected mineral shock. Power rationing forced the temporary shutdown of magnesium, silicon metal, and rare earth processing facilities. Magnesium is a crucial component in aluminum alloys used in the automotive and aerospace sectors. Because China produces the majority of global magnesium, the shutdown triggered warnings of potential production halts for European industries. This event demonstrated that even non-geopolitical disruptions—such as domestic energy constraints—could trigger global supply chain crises when industries rely on highly concentrated production.

The geopolitical landscape shifted dramatically in 2022 with Russia's full-scale invasion of Ukraine. Western sanctions targeted Russian financial institutions, energy exports, and industrial goods, while Russia responded with its own restrictions. One of the most significant disruptions involved noble gases—neon, argon, and krypton—used in semiconductor manufacturing. Prior to the conflict, much of the world's neon supply came from Ukraine, sourced from Russian steel byproducts and purified in Ukrainian facilities. The war severed these production chains, forcing semiconductor companies to seek alternative sources in Japan, South Korea, and the United States. This highlighted the hidden dependencies that connect raw materials, specialized processing, and advanced technology.

The war also triggered fears that Russia might restrict exports of palladium or titanium, both essential for global industries. Although Russia refrained from weaponizing these minerals fully, the risk underscored the strategic vulnerability of nations dependent on adversarial suppliers. Western companies accelerated efforts to diversify away from Russia, reshaping procurement patterns in aerospace, automotive, and electronics sectors.

In 2023, another major flashpoint emerged when China implemented export licensing requirements for gallium and germanium. These materials are crucial for compound semiconductors, infrared optics, and advanced electronic components. China's restrictions were widely interpreted as a response to U.S. and allied export controls on advanced chips and lithography equipment. The new licensing rules immediately created delays, confusion, and price volatility. Western firms scrambled to source alternative supplies, but global production outside China remained limited. This move crystallized the reality that China could—and would—use minerals as geopolitical leverage in technology competition.

That same year, China imposed restrictions on certain graphite products, essential for lithium-ion battery anodes. Because virtually all battery-grade graphite is processed in China, this policy sent shockwaves through EV supply chains. It also signaled that China was willing to expand mineral leverage beyond niche materials into large-volume markets crucial to global electrification.

Meanwhile, 2023 and 2024 witnessed disruptions in the Democratic Republic of Congo's cobalt sector due to political tension, regulatory uncertainty, and intermittent violence. These disruptions affected global cobalt supply chains, especially for EV batteries. The instability reinforced concerns about the risks associated with sourcing from politically fragile regions and the need for diversification and responsible mining practices.

By 2024 and early 2025, tensions between the U.S. and China escalated further with new export controls, countersanctions, investment screening measures, and restrictions on advanced semiconductor technology. Critical minerals were central in this confrontation. Each new measure triggered responses from industry, governments, and investors, reinforcing the perception that minerals are no longer neutral commodities but instruments of strategic competition.

Taken together, the flashpoints of 2010–2025 map a clear trajectory: increasing concentration of supply, rising geopolitical competition, expanding resource nationalism, and the frequent use of minerals as tools of coercion or leverage. They reveal a world in which mineral security is inseparable from technological and national security. They also demonstrate that the patterns identified in earlier chapters are not theoretical—they have already shaped global politics and will continue to do so.

States are increasingly treating critical materials as strategic assets to be buffered against uncertainty. Stockpiles, national reserves, and preemptive safeguards reflect a shift from market reliance toward risk management. In an environment of recurring disruption, preparedness itself becomes a form of power.

CHAPTER 35 – STRATEGIC STOCKPILES AND NATIONAL SECURITY MEASURES

As critical minerals become the backbone of energy systems, digital infrastructure, transportation, defense, and advanced manufacturing, nations are rediscovering an old tool with new urgency: the strategic stockpile. In earlier eras, stockpiles primarily focused on oil, grains, or wartime materials. Today, governments recognize that minerals once considered obscure—gallium, germanium, lithium, cobalt, rare earths, graphite, titanium, palladium, nickel, and many others—can determine national security outcomes. Modern economies cannot function without them, and disruptions—whether caused by conflict, sanctions, natural disasters, or political decisions—can quickly cascade into industrial paralysis. Strategic stockpiles therefore represent both a defensive measure and a proactive instrument of state strategy.

The logic behind stockpiling is straightforward: maintain an emergency reserve of critical materials to ensure national resilience during periods of supply disruption. But the implementation is far more complex. Stockpiles must consider fluctuating demand, evolving technology, geopolitical risk, and the uneven speed of industrial transformation. A nation might stockpile cobalt, only to find that battery chemistries evolve to reduce cobalt content. Or it might focus on rare earth oxides while the real bottleneck exists in magnet manufacturing capacity. Effective stockpiling requires not only mineral reserves but strategic foresight.

The United States provides a clear example of this evolution. The National Defense Stockpile (NDS), originally established during World War II, at one point held hundreds of thousands of tonnes of materials—from chromium and manganese to rubber and tin. As geopolitical tensions eased in the late twentieth century, the stockpile was downsized, and during the 1990s large quantities were sold off to reduce costs. But by the 2010s, it became apparent that the U.S. had underestimated its vulnerability. The rare earth crisis of 2010, dependence on foreign cobalt and lithium, and China's dominance in refining prompted a reevaluation of stockpile policy. The NDS has since been modernized to include rare earth elements, battery minerals, and specialty alloys used in defense platforms. Congress has expanded funding, directed new procurements, and authorized the use of emergency authorities to support domestic production of critical minerals.

Yet even the U.S. recognizes that strategic stockpiling cannot solve supply chain fragility alone. Stockpiles must be integrated with broader national security measures: investment in domestic refining, public-private partnerships, recycling programs, trade alliances, and support for alternative supply chains. The United States increasingly treats stockpiling as one tool among many in a comprehensive mineral security strategy—not as a standalone solution.

China, by contrast, has built one of the world's most sophisticated and opaque strategic mineral systems. Through state reserves, industrial stockpiles, and company-level inventories, China can influence market conditions, stabilize domestic industries, or tighten global supply. China has accumulated reserves of rare earths, cobalt, nickel, tungsten, and other minerals to buffer its manufacturing base against shocks. Its stockpiling is not purely defensive; it also creates market leverage. By adjusting release schedules or expanding purchases, China can influence prices and affect the economics of competing projects abroad. This flexibility gives Beijing a powerful instrument to shape global mineral markets.

Japan offers another instructive model. Following the 2010 rare earth incident, Japan rapidly expanded its stockpiling of rare earths, tungsten, cobalt, and other critical materials. Its approach is coordinated through the Japan Oil, Gas and Metals National Corporation (JOGMEC), which maintains reserves and facilitates joint stockpiling between government and private industry. This hybrid approach ensures that Japan's advanced manufacturing sectors—especially automotive and electronics—have access to emergency supplies without disrupting markets. Japan's stockpile strategy is supported by diversification efforts, including mining partnerships in Vietnam, Australia, and Africa, and investments in recycling technologies.

South Korea has pursued a similar path, using the Korea Mine Rehabilitation and Mineral Resources Corporation to secure materials essential for semiconductors, batteries, and machinery. Given its heavy industrial concentration and dependence on mineral imports, South Korea integrates stockpiling with active diplomatic outreach and corporate procurement strategies. Its stockpiles are designed not only to protect against geopolitical risk but to stabilize domestic manufacturing during global market volatility.

Europe's approach to strategic stockpiling is more fragmented but rapidly evolving. The European Union has encouraged member states to coordinate reserves, particularly for lithium, cobalt, rare earths, and magnesium. However, because industrial needs differ across member states, Europe relies more heavily on diversified supply chains, recycling, and industrial alliances than on centralized stockpiles. Still, individual states—such as France and Germany—maintain their own strategic reserves of critical materials for defense and industrial resilience. The EU's Critical Raw Materials Act now formalizes stockpiling as part of a broader resource security strategy, including monitoring mechanisms, early-warning systems, and emergency coordination protocols.

Beyond major powers, many emerging economies are beginning to consider strategic stockpiles. India has launched initiatives to establish reserves of lithium, cobalt, and nickel while simultaneously investing in overseas mining partnerships. Gulf countries, particularly Saudi Arabia and the UAE, are exploring stockpiling as part of their broader economic diversification strategies. Even smaller nations with advanced manufacturing sectors—such as Singapore or Taiwan—are assessing their vulnerabilities in semiconductor materials, gases, and specialty metals.

Stockpiles, however, face several inherent challenges. First is the question of which materials to stockpile. Technology evolves rapidly, and strategic minerals today may become less important tomorrow. For example, if sodium-ion batteries become widespread, lithium stockpiles may lose strategic value. If rare-earth-free magnet technologies become commercially viable, the need for neodymium, praseodymium, and dysprosium may decline. Governments must anticipate future demand trends, not merely respond to present shortages.

Second is the issue of scale. Some minerals are used in such large quantities—copper, nickel, manganese—that stockpiling meaningful reserves becomes financially and logistically impractical. Others, such as gallium or germanium, are used in small quantities but require extremely high purity, making storage more technically complex.

Third, stockpiles risk distorting markets if mismanaged. Large, sudden purchases can cause price spikes or inadvertently benefit competitors. Premature releases may undermine domestic producers or send misleading signals to global markets. Successful stockpile management requires coordination, transparency (when appropriate), and market sensitivity.

Fourth, stockpiling alone does not address upstream vulnerabilities. For materials where supply chains are dominated by a handful of countries, such as graphite or rare earth magnets, stockpiles provide only temporary relief. Long-term security requires building domestic or allied processing capacity, developing substitutes, and investing in recycling.

Fifth, stockpiles are expensive. Many critical minerals—like iridium, palladium, scandium, or high-purity germanium—are expensive to acquire and store. Budgetary constraints can limit stockpiling ambitions, particularly for countries with competing priorities in energy, defense, or public welfare.

However, despite these challenges, strategic stockpiles remain indispensable tools for national resilience. They serve as buffers during crises, deterrents against coercion, and stabilizers for key industries. They also complement broader national security strategies. In a world where supply chains are increasingly weaponized, stockpiles reduce vulnerability and increase leverage.

The future of strategic stockpiles is trending toward greater sophistication. Automated monitoring systems, real-time inventory tracking, and predictive demand modeling will help governments adjust reserves dynamically. Joint stockpiling—where allied countries coordinate reserves for shared materials—could become more common, especially among NATO members, Quad partners, and EU states. Private–public stockpile models may expand, allowing industries to maintain reserves subsidized by government support. Digital twins of supply chains may enable more accurate forecasting of mineral needs across sectors.

In the broader geopolitical picture, stockpiles will play a crucial role in the emerging era of mineral competition. They will support energy transition systems, stabilize semiconductor manufacturing, ensure military readiness, and underpin industrial transformation. As the global scramble for critical minerals intensifies, nations that manage stockpiles effectively—alongside mining, refining, alliances, and innovation—will be better positioned to navigate uncertainty.

This concludes Part IV. The next section of the book will shift from geopolitics and conflict toward long-term scenarios, sustainability pathways, and strategic recommendations—addressing how the world can build a secure, equitable, and resilient mineral future in an era of unprecedented technological change.

PART V – MARKETS, ACTORS, AND GLOBAL DYNAMICS
CHAPTER 36 – MINING GIANTS AND STATE-OWNED ENTERPRISES

The global critical minerals industry rests on the shoulders of a relatively small number of extremely powerful mining companies and state-controlled enterprises. These actors shape extraction capacity, influence geopolitical alignments, determine market stability, and set the pace for industrial expansion in the clean-energy and digital sectors. Mining giants—both private multinationals and state-owned enterprises—control the flow of lithium, cobalt, nickel, rare earths, graphite, copper, uranium, and dozens of specialized metals that underpin the global economy. Their strategic decisions reverberate far beyond the mining sector, influencing everything from electric vehicle production to national security strategy.

Critical minerals differ from bulk commodities like iron ore or coal in one important way: their supply chains are far more concentrated. A handful of companies dominate production of minerals that have no near-term substitutes and are difficult to source elsewhere. This concentration creates structural power. It gives mining companies the ability to shape global markets, and it gives states—especially those with strong state-owned enterprises—direct influence over strategic resources.

Historically, the world of mining has been led by a set of large, diversified corporations: BHP, Rio Tinto, Vale, Glencore, and Anglo American. These firms built their empires on iron ore, copper, coal, nickel, and other high-volume minerals. But as demand for lithium, cobalt, rare earths, graphite, and other specialty materials skyrocketed, many of these giants began shifting their portfolios toward the minerals of the energy transition. Some moved cautiously. Others accelerated rapidly, especially when it became clear that renewable energy, batteries, and electronics would reconfigure global mineral demand for decades to come.

Glencore stands out as one of the most influential players in the critical minerals supply chain. Known as both a miner and one of the world's most powerful commodity traders, Glencore controls major cobalt and nickel assets in the Democratic Republic of Congo, Canada, and Australia. Its Katanga and Mutanda operations form the backbone of global cobalt supply, feeding battery manufacturers and electronics companies around the world. Because cobalt is both geographically concentrated and indispensable to many battery chemistries, Glencore's position gives it significant leverage. It has also produced germanium and indium as byproducts from its zinc operations, making it relevant across multiple niche markets. Glencore's ability to negotiate directly with automakers and battery companies marks a shift in the industry: mining companies increasingly interact directly with end users, no longer merely supplying metals but shaping entire supply chain strategies.

Rio Tinto is another major actor reshaping its portfolio. Traditionally known for iron ore and aluminum, it has expanded into critical minerals with a mixed degree of success.

The company produces borates in California, titanium feedstocks in South Africa and Madagascar, and is developing lithium assets such as the Jadar project in Serbia. Though the latter faced strong environmental opposition and remains stalled, Rio Tinto's scale, technical expertise, and logistical network position it to play a much larger role in critical minerals if it chooses. The company also owns rare-earth-rich deposits in Tanzania and has launched efforts to extract tellurium as a byproduct of its copper operations in Utah. As demand rises, Rio Tinto's access to capital and engineering capabilities may allow it to bring new projects online faster than smaller competitors.

Vale, historically rooted in iron ore and nickel, has become central to the future of energy transition minerals. Its Indonesian nickel operations, together with Canadian and Brazilian assets, feed the global battery industry. Nickel has become a geopolitical battleground as high-nickel cathode chemistries rise to prominence. Vale's partnerships with automakers and battery manufacturers—combined with Indonesia's aggressive industrial policy—place it at the heart of EV supply chains. Meanwhile, Vale's shift toward cleaner mining practices aligns with global ESG expectations, allowing it to remain a preferred supplier for Western industries seeking ethically sourced materials.

Beyond private multinationals, state-owned enterprises (SOEs)—particularly Chinese SOEs—are among the most powerful actors in the critical minerals landscape. China's industrial strategy has been deliberate, coordinated, and long-term. Through its SOEs, China built near-total dominance in rare earths, cobalt refining, graphite processing, and a substantial share of global lithium and nickel refining. These companies are deeply integrated with state objectives, receiving financing from policy banks, diplomatic support, and geopolitical backing through the Belt and Road Initiative.

China Northern Rare Earth Group is the largest rare earth producer in the world, anchored around the Bayan Obo deposit in Inner Mongolia. Alongside Minmetals, Chinalco, and the newly formed China Rare Earth Group, these SOEs have unparalleled command over global rare earth mining and separation. Their ability to increase or decrease output, invest in new operations abroad, and shape pricing through coordinated policy makes them critical players not only in industry but in geopolitics. Because rare earths are essential for EV motors, wind turbines, data centers, and military platforms, the dominance of Chinese SOEs represents a structural vulnerability for many Western nations.

China Molybdenum Co. (CMOC) provides another example of the strategic reach of Chinese SOEs. Its acquisition of the Tenke Fungurume mine in the DRC—one of the world's largest cobalt and copper deposits—gave China direct control over one of the most important battery mineral assets on the planet. With state backing, CMOC expanded production rapidly, enabling China to maintain its role as the central hub in global cobalt refining, even though the raw materials originate in Africa. CMOC's approach—combining foreign asset purchases with domestic refining capacity—symbolizes the core of China's minerals strategy.

Zijin Mining, though nominally private, operates with similar state-aligned ambitions. It has aggressively acquired lithium, gold, copper, and rare earth assets across Latin America, Africa, and Central Asia. Its stake in the Kamoa-Kakula copper mine and various lithium projects demonstrates China's ambition to secure upstream resources globally. Zijin's expansion mirrors Beijing's understanding that control over raw materials is inseparable from control over future technologies.

In other parts of the world, state-owned enterprises play different roles. Kazakhstan's Kazatomprom dominates the global uranium market, supplying fuel for nuclear power at a time when many countries reassess nuclear's role in energy security. Russia's Rosatom, while primarily a nuclear energy conglomerate, has sought stakes in lithium projects to diversify its strategic position. In the Middle East, Saudi Arabia's Ma'aden and the broader Saudi Public Investment Fund have begun moving into critical minerals, aiming to position the Kingdom as a future hub for battery materials and green metals. These investments reflect a global shift in strategic thinking: nations rich in capital but poor in critical minerals seek to secure supply by acquiring overseas assets.

Alongside these giants, a new class of specialized companies is emerging. Lynas Rare Earths of Australia remains the largest rare earth producer outside China, operating the Mt Weld mine and processing facilities in Malaysia and expanding into Australia and the United States. MP Materials in the United States has revived the Mountain Pass mine and is reshoring rare earth separation and magnet production. These firms differ from traditional mining giants: they are vertically integrating from mine to magnet, driven by national security incentives and direct partnerships with automakers and defense companies. Their growth illustrates how shifts in geopolitics and industrial policy are creating opportunities for specialized players.

In lithium, companies such as Albemarle, SQM, Ganfeng, Tianqi, Pilbara Minerals, and Liontown Resources form a dynamic ecosystem of global producers. Albemarle and SQM dominate the Atacama brine fields, while Ganfeng and Tianqi have aggressively expanded into joint ventures in Australia, Latin America, and China. Their control over brines, spodumene, and chemical conversion plants makes them central actors in the battery revolution.

Nickel is shaped by a mix of actors: Vale, Norilsk Nickel, BHP, Indonesian SOEs, and Chinese-invested ventures dominate global supply. Indonesia, supported by partnerships with Chinese firms, has become the world's fastest-growing nickel powerhouse through high-pressure acid leach (HPAL) plants that produce battery-grade materials. The emerging Indonesia–China axis in nickel is now one of the most strategic supply chains in the world.

Graphite supply is dominated by Chinese companies such as BTR and Shanshan, while Syrah Resources in Mozambique and the United States is building alternative sources with Western support. Platinum group metals remain concentrated in South African companies like Anglo American Platinum and Impala Platinum, although diversification efforts are underway.

Collectively, these actors—private giants, state-owned enterprises, and specialized producers—form the backbone of global mineral supply. Their decisions on investment, expansion, pricing, and partnerships will shape the next decades of industrial development. Mining giants can accelerate or slow the energy transition by deciding which projects to fund. SOEs can reinforce geopolitical alliances or deepen dependencies. Specialized firms can provide alternatives or create new chokepoints.

Value in critical materials is created downstream, where refining, chemical processing, and component manufacturing take place. These industries convert raw minerals into the high-performance inputs required by batteries, electronics, and renewable energy systems. Control over this segment determines who captures technological and economic advantage.

CHAPTER 37 – REFINERS, CHEMICAL PRODUCERS, AND MAGNET MAKERS

If mining giants sit at the beginning of the critical minerals value chain, refiners, chemical processors, and magnet manufacturers form its decisive center of gravity. Their role is not simply to convert ore into usable materials; they determine where value is captured, which nations hold leverage, and how industrial ecosystems evolve. In the world of critical minerals, mining is often easier than refining. Many countries can dig minerals out of the ground, but only a handful can transform them into high-purity chemicals or advanced components. This chemical and metallurgical bottleneck shapes the geopolitics, economics, and technological direction of the twenty-first century.

Across the entire supply chain—from cobalt sulphate to lithium hydroxide, from nickel matte to rare earth oxides, from gallium metal to neodymium-iron-boron magnets—refining and processing are where most of the value is added. This is also where strategic vulnerabilities are most acute. As noted in the research base, China processes more than half of the world's lithium, two-thirds of its cobalt, one-third of its nickel, and nearly all rare earth elements . This dominance did not emerge by chance; it reflects decades of strategic policy, infrastructure investment, and a willingness to operate under environmental and regulatory conditions that many competitors found uneconomic.

Refiners are the true "chokepoint" actors. They determine purity, capacity, and availability of battery-grade, magnet-grade, and semiconductor-grade materials. Chemical producers further refine these materials into cathode powders, anode materials, or high-purity oxides. Magnet makers, at the end of this chain, translate rare earth metals into the highest-value components used in electric vehicles, wind turbines, robotics, and defense systems.

Together, these three groups hold the industrial world in a tight interdependency.

The global refining landscape is dominated by China, whose processing ecosystem integrates dozens of state-owned and private firms. Companies like Jinchuan Group, one of the world's most important cobalt refiners, import cobalt from the Democratic Republic of the Congo and convert it into battery-grade chemicals used by China's cathode and battery manufacturers . China's processing advantage extends to lithium as well: firms like Ganfeng Lithium and Tianqi Lithium own stakes in Australian and South American mines, but their real power lies in conversion capacity inside China. Even when lithium is mined in Australia or Chile, it is often shipped to China for refining.

Nickel processing follows a similar pattern. Indonesia produces vast quantities of nickel ore, yet historically it lacked refining capacity, sending raw ore to China for processing. Only recently—driven by Indonesian resource nationalism and Chinese investment—have large high-pressure acid leach (HPAL) plants been built locally. Still, China remains the principal hub where nickel matte and mixed hydroxide precipitate are transformed into battery-grade nickel sulphate.

Refining is even more concentrated for rare earths. Solvent extraction plants—complex facilities requiring thousands of tanks—have long been clustered in Inner Mongolia and southern Chinese provinces. It is in these facilities that companies like China Northern Rare Earth and the newly consolidated China Rare Earth Group separate mixed rare earth concentrates into high-purity oxides such as neodymium, praseodymium, dysprosium, and terbium. The research highlights how deeply rooted this industrial capacity is, built through decades of state-supported expansion and know-how accumulation .

Outside China, refining activity is limited but growing. In Europe, companies like Umicore and BASF are world-class producers of battery cathode materials, dependent on long-term cobalt and nickel supply agreements. Umicore also operates one of the world's most sophisticated metallurgical recycling hubs, which processes cobalt, nickel, palladium, and other high-value materials from spent batteries and industrial scrap . In Finland, Umicore's acquisition of Freeport Cobalt created one of the few major cobalt refining facilities outside China.

Japanese chemical producers, long leaders in high-purity materials for electronics and automotive applications, supply advanced cathode materials, metal powders, and precision oxides. South Korean companies such as LG Chem and POSCO Chemical have invested heavily in nickel, cobalt, and lithium precursors, building a vertically integrated battery ecosystem that relies on chemical processing expertise.

Still, these global efforts pale in comparison to China's integrated system. China not only refines the majority of key minerals but also houses the world's most extensive network of magnet manufacturers—a sector that remains one of the most strategically sensitive in global markets.

Magnet production—especially neodymium-iron-boron (NdFeB) magnets—represents the apex of value in the rare earth supply chain. Magnets are essential for EV motors, wind turbines, robotics, industrial machinery, consumer electronics, and high-precision defense systems. This is why "magnets are where the money is," as the research notes, and why China produces around 90 percent of NdFeB magnets worldwide .

China's dominance is so complete that even if Western nations mine or refine rare earths, they still often send materials to China for magnet production. Japanese firms, especially Shin-Etsu Chemical, remain world-class magnet producers, but their market share is modest compared to China's.

Outside Asia, the magnet industry is relatively small but politically critical. The United States, through MP Materials, is rebuilding a domestic magnet supply chain, including manufacturing facilities for NdFeB alloys and finished magnets. Europe is investing in magnet capacity in Germany and the UK, but environmental constraints, high energy costs, and limited industrial clusters slow progress. Magnet manufacturing requires precision metallurgy, controlled atmospheres, and specialized sintering equipment—capabilities that take years to develop.

Cobalt, lithium, nickel, and rare earth chemical processors shape global industrial strategy by supplying the materials required for battery and motor technologies. Battery-grade chemicals—lithium hydroxide, lithium carbonate, nickel sulphate, cobalt sulphate, manganese sulphate—must meet strict purity standards to ensure performance and safety. High-purity graphite used in battery anodes must be processed through energy-intensive techniques that China controls almost entirely.

At the frontier of processing are refiners of gallium and germanium, two materials essential for compound semiconductors, infrared optics, and high-frequency electronics. These metals are typically recovered as byproducts—gallium from bauxite refining and germanium from zinc or coal processing . China dominates these niche refining sectors as well, and its 2023 export controls on gallium and germanium revealed how strategic this dominance has become.

Refiners and chemical processors therefore operate at the intersection of minerals, technology, and geopolitics. Their decisions on output, export, and investment directly influence EV production, renewable energy deployment, semiconductor manufacturing, and defense readiness. For instance, when China introduced licensing requirements for graphite exports in 2023, global EV manufacturers immediately faced uncertainty because nearly all spherical purified graphite for battery anodes is processed in China .

Magnet makers create the final products with the highest strategic value. A single EV motor may contain several kilograms of high-performance magnets; a single offshore wind turbine may require hundreds of kilograms. Defense systems—from radar arrays to precision-guided munitions—depend on rare earth magnets. In this context, the research highlights China's consideration of export controls on magnet technology itself, pointing to the next phase of strategic competition: not just controlling materials, but controlling the manufacturing know-how that turns those materials into irreplaceable components .

The fragility of refining and magnet supply chains reflects several structural realities. First, the chemical complexity of refining creates barriers to entry. Solvent extraction for rare earths requires handling thousands of stages and managing radioactive byproducts. Nickel and cobalt refining involve high-pressure, corrosive chemistries. Lithium refining requires precise control of impurities to achieve battery-grade quality. Second, economies of scale—especially those enjoyed by China—make new entrants struggle to compete on cost. Third, environmental regulations in Western countries increase costs and permitting timelines, pushing processing to jurisdictions with looser oversight.

Yet diversification is underway. New rare earth separation facilities are being built in Australia, the United States, and Canada. Lithium refineries are expanding in Australia and Europe. Nickel processing capacity is growing in Indonesia. Japan and Korea continue to advance their chemical expertise. Magnet manufacturing initiatives in the U.S., Europe, and India reflect growing recognition of the need for domestic capability.

Refiners, chemical producers, and magnet makers determine whether critical minerals are simply raw resources or the backbone of technological power. They are the link between geology and industry, between extraction and innovation. Their central role makes them indispensable to the future of the energy transition and the global economy.

Automakers and technology firms are moving upstream to manage material risk. Vertical integration, long-term offtake agreements, and direct investment in supply chains are reshaping competitive dynamics across clean-energy industries. In this environment, access to materials becomes as decisive as innovation itself.

CHAPTER 38 – AUTOMAKERS, TECH CORPORATIONS, AND VERTICAL INTEGRATION

The accelerating global transition toward electrification, digitalization, and automation has pushed automakers and major technology corporations into a position they never occupied in previous industrial eras: they are now among the world's most influential consumers of critical minerals. Their dependence on lithium, nickel, cobalt, graphite, rare earths, copper, gallium, germanium, and dozens of other materials has transformed them from downstream manufacturers into strategic actors deeply involved in mining, refining, and chemical processing. Electrification is forcing these companies to rethink the boundaries of their business models. Vertical integration—long considered outdated in many industries—has become a central strategic imperative.

For more than a century, automakers operated with a predictable supply chain model: mining companies extracted raw materials, refineries processed them, chemical manufacturers transformed them, and component suppliers delivered finished parts. Carmakers then assembled vehicles, focusing on engineering, branding, and consumer experience. This model no longer works in an era where the cost, availability, and purity of minerals determine production viability. Battery materials shortages can idle entire manufacturing lines, while disruptions in magnet supply can halt EV motor production overnight. The new reality is clear: automakers can no longer afford to be passive buyers—they must become active participants across the entire materials pipeline.

Tesla recognized this shift earlier than most. From its earliest days, the company understood that the battery supply chain was not simply a procurement problem but a strategic vulnerability. It invested heavily in long-term supply contracts for lithium, nickel, and cobalt while simultaneously partnering with cathode and anode manufacturers to secure chemical inputs. It announced plans to mine lithium in the United States, explored refining facilities in Texas, and developed proprietary battery chemistries to reduce exposure to volatile cobalt markets. Although Tesla is not a mining company, its direct involvement in upstream materials has forced the rest of the automotive industry to rethink traditional supply chain boundaries.

Other automakers have followed suit. General Motors has signed agreements with lithium developers in the United States, South America, and Australia, and has invested in battery precursor manufacturing. GM is also partnering on cathode material production and has explored joint ventures for rare earth magnet manufacturing—an unprecedented move for a company historically focused on final assembly. Ford has pursued similar strategies, building relationships with nickel and lithium suppliers and entering partnerships with battery giants to ensure stable access to materials. These efforts reflect a recognition that the security of critical minerals is inseparable from corporate competitiveness.

European automakers, facing stricter environmental regulations and carbon-neutrality mandates, have intensified their vertical integration strategies. Volkswagen announced the creation of its own battery division, PowerCo, which aims to build gigafactories across Europe and secure raw materials through direct offtake agreements. BMW and Mercedes-Benz have invested in cobalt and nickel supply chains that emphasize traceability and ESG compliance. Renault has partnered on cathode materials and recycling facilities. Stellantis has entered agreements for lithium supply in North America and Europe while simultaneously investing in battery joint ventures. These companies no longer view mineral sourcing as a procurement task but as a strategic dimension of their survival in a competitive market increasingly shaped by China's industrial dominance.

Vertical integration is not limited to automotive companies. Technology corporations—especially those involved in consumer electronics, cloud infrastructure, and semiconductors—are becoming deeply engaged in critical mineral strategies. Apple, for instance, has sought to secure cobalt and rare earth supplies through direct sourcing agreements, bypassing traditional intermediaries. Though Apple does not mine or refine materials, it imposes strict standards on suppliers, shaping global practices around traceability, environmental compliance, and ethical sourcing. Its influence on the cobalt sector, given its enormous demand for rechargeable batteries, is particularly significant.

Semiconductor companies also play an expanding role. They depend on dozens of specialty materials—high-purity silicon, gallium, germanium, tungsten, indium, tantalum—and are increasingly aware of the geopolitical fragility of these supply chains. Companies such as TSMC, Samsung, and Intel do not mine these materials, but they work closely with chemical suppliers to ensure purity levels and supply continuity. Their input influences national policies and international agreements, especially in regions where semiconductor manufacturing is a pillar of national security. These companies operate in a world where a shortage of gallium or neon can disrupt global electronics production, making them indirect actors in mineral geopolitics.

Cloud providers and data center operators—Amazon, Google, Microsoft—have become major stakeholders in the copper and rare earth value chains due to the enormous scale of their server farms, cooling systems, and power infrastructure. Their long-term planning horizons have pushed them to advocate for resilient electrical equipment supply chains and to support alternative materials research for power electronics and magnet technologies. They are not yet directly involved in mining or refining, but their influence on demand patterns is reshaping investment priorities.

Vertical integration increasingly takes the form of joint ventures, long-term offtake agreements, and direct investments. Instead of owning mines outright, automakers and tech corporations partner with mining companies to secure future supply. These agreements often span decades and include financial commitments, technology sharing, and ESG requirements. Automakers may provide early-stage funding to mining projects in exchange for guaranteed material supply. In some cases, they invest directly in refining facilities to reduce dependence on foreign processing hubs.

The battery industry is where vertical integration has developed most dramatically. Companies like CATL, BYD, LG Energy Solution, Panasonic, and Samsung SDI are building vertically integrated ecosystems that encompass materials sourcing, precursor production, chemical refining, cell manufacturing, battery recycling, and in some cases vehicle assembly. Chinese companies in particular have embraced full-spectrum integration. BYD mines lithium, refines it, manufactures batteries, builds motors, makes electronics, and assembles vehicles—an industrial model reminiscent of early 20th-century conglomerates but optimized for modern electrification.

This integration gives Chinese firms enormous strategic advantages. They can manage cost volatility, respond quickly to supply disruptions, and maintain control over key technologies. It is no coincidence that China leads in global EV sales, battery production, and renewable energy deployment. Its corporations operate across every layer of the mineral and manufacturing stack, supported by state incentives, infrastructure, and long-term strategic planning.

Western corporations, by contrast, are still catching up. Many rely heavily on suppliers in Asia, especially for cathode materials, anode materials, separators, and battery-grade chemicals. This dependency constrains their geopolitical leverage. The more upstream the supply chain moves toward China, the more difficult it becomes for Western firms to ensure supply independence. Vertical integration efforts by U.S. and European automakers therefore aim to shift the balance gradually, but these efforts face challenges such as limited refining capacity, high energy costs, permitting delays, and complex environmental requirements.

Recycling is becoming another pillar of vertical integration. Automakers and tech companies increasingly view battery recycling not only as an ESG requirement but as a strategic resource stream. By recovering nickel, cobalt, lithium, and manganese from spent batteries, they reduce the need for new mining and protect themselves from geopolitical risks. Tesla, Volkswagen, BMW, and Apple all have recycling partnerships or in-house programs to close the loop. Vertical integration thus extends beyond production to end-of-life management, reflecting a larger shift toward circular industrial systems.

Vertical integration also affects market dynamics in unexpected ways. As automakers secure long-term supply, traditional commodity traders lose some influence. As tech companies invest in refining partnerships, chemical producers face pressure to adapt to new performance standards. As Chinese firms dominate integrated ecosystems, Western companies face higher capital requirements and greater competitive pressure.

The shift toward vertical integration suggests a reconfiguration of global industrial power. Automakers and tech corporations are no longer competing merely on design, branding, or software—they are competing on access to minerals, command over refining technologies, control of intellectual property for battery chemistries, and resilience of supply chains. The winners of this race will be those who master the full spectrum of materials, manufacturing, and market strategy.

Between producers and end users, traders and financial intermediaries exert quiet but decisive influence. Their control over price discovery, liquidity, and logistics shapes volatility as much as physical supply. In critical minerals markets, financial architecture becomes a strategic layer of power.

CHAPTER 39 – TRADERS, FINANCIALIZATION, AND PRICE VOLATILITY

Critical minerals do not move from mines to factories in a straight line. They flow through a complex network of commodity traders, financial intermediaries, brokers, exchange platforms, opaque specialty markets, and informal trading routes. These actors—often hidden from public view—are among the most influential in shaping prices, liquidity, market stability, and even the pace of industrial expansion. Their decisions determine whether a factory receives cobalt next month, whether lithium prices spike or collapse, or whether a geopolitical shock reverberates through global supply chains. In the world of critical minerals, the power of traders and financial markets rivals that of mining giants or state-owned enterprises.

Trading houses occupy a unique strategic position. Firms such as Glencore and Trafigura act as the circulatory system of the mineral economy, linking mines in the Democratic Republic of Congo or Indonesia with battery manufacturers in China, Europe, Korea, and the United States. Their ability to stockpile, blend, ship, and arbitrage materials gives them substantial influence over market conditions. These traders operate across borders, time zones, and regulatory regimes, navigating volatile markets with a sophistication few other industries possess.

For some minerals—like copper, nickel, or tin—formal exchanges such as the London Metal Exchange (LME) provide futures contracts, price benchmarks, and standardized trading structures. But for many critical minerals, no such infrastructure exists. Rare earths, gallium, germanium, high-purity graphite, and many lithium chemicals are traded in opaque, bilateral contracts, with prices reported by specialized agencies rather than discovered in open markets. This lack of transparency creates fertile ground for volatility and speculation. A single large purchase order, a shipping delay, or a policy announcement can shift prices sharply because the market lacks depth and liquidity.

The LME's experience with nickel in March 2022 highlighted the risks of thin liquidity even in "exchange-traded" markets. When a major Chinese producer's short position collided with a sudden price surge, nickel prices spiked by more than 250 percent in two days, forcing the LME to halt trading and cancel transactions. This episode eroded confidence in the LME's ability to manage extreme events and exposed how vulnerable critical minerals can be to financial shocks . It also demonstrated how a single actor's position—especially if leveraged or opaque—can destabilize global markets.

Outside the LME, speculation occurs in more informal ways. Many critical minerals experience prolonged boom-and-bust cycles because supply cannot ramp up quickly. It can take seven to ten years to bring a new mine into production, while demand can shift dramatically in a few quarters. Lithium illustrates this dynamic perfectly. Lithium carbonate prices climbed from around $6,000 per ton in 2020 to nearly $70,000 in late 2022, driven by EV demand and perceived scarcity, before collapsing again by mid-2023 as new supply came online and sentiment cooled . Cobalt, too, has swung wildly: rising on EV hype in 2017–18, collapsing on oversupply in 2019, surging again in 2021, then falling in 2023 as market expectations shifted .

Such volatility does more than affect balance sheets. It discourages investment at precisely the moments when new capacity is needed. Financiers remember the Molycorp rare earth bubble of the early 2010s—prices spiked after China's export restrictions, capital poured in, mines reopened, and then prices collapsed before new projects could stabilize. Molycorp went bankrupt. This pattern haunts investors today: no one wants to fund a lithium or rare earth project at the top of the cycle only to see prices crash when production begins. Because supply lags demand by years, speculative swings can undermine long-term planning in exactly the materials where planning matters most.

To manage these cycles, major traders often act as stabilizers—or destabilizers. Glencore, for instance, has at times withdrawn cobalt from the market by stockpiling material, smoothing price cycles or positioning itself strategically. Trafigura has financed stockpiles of artisanal cobalt in the DRC, purchasing material that might otherwise flood the market and depress prices . In another case, a Chinese consortium supported a fund (Cobalt 27) designed to hold physical cobalt as an investment asset, effectively turning metal into a financial instrument. These activities blur the line between commodity trading and financial speculation.

The growth of "green metals funds" has deepened financialization. As investors seek exposure to the expected boom in EVs, renewables, and semiconductors, capital has flowed into mining juniors—small exploration companies listed on stock exchanges in Canada, Australia, and the UK. Their valuations often track sentiment rather than fundamentals, rising on news of EV demand or policy announcements, and falling when macroeconomic fears arise. Many of these companies will never produce a single ton of material, but their share prices move markets and shape expectations. Financialization thus amplifies both optimism and panic.

The role of governments in financialized markets is growing. China's State Reserve Bureau regularly intervenes by purchasing cobalt, rare earths, or other strategic materials when prices are low and releasing them when domestic industries need price relief . These interventions serve dual goals: stabilizing domestic manufacturers and reinforcing China's position as a central actor in mineral markets. When China consolidated rare earth producers into the China Rare Earth Group in 2021, it effectively formed a state-guided cartel with the ability to coordinate output and influence prices globally. Market observers noted how such structural consolidation mirrors OPEC's function—even if no formal cartel exists.

In contrast, Western governments tend not to intervene directly in prices but influence markets through policy signals, subsidies, or support for new projects. When the U.S. Department of Energy or European governments announce funding for critical mineral projects, investor interest rises. The Inflation Reduction Act's EV mineral sourcing rules reshaped lithium and nickel investment flows, pulling capital toward "friend-shored" jurisdictions. This reorientation shows how policy can influence financial markets without direct price controls.

Another layer of volatility arises from illegal, informal, or political supply channels. Rare earths from Myanmar—often mined informally and smuggled across the border into China—have caused price fluctuations when political instability disrupted flows. Crackdowns in Myanmar in 2021 temporarily drove dysprosium prices up by roughly 30 percent before trade resumed . Artisanal cobalt from the DRC introduces similar unpredictability. When authorities intervene, supply tightens; when artisanal output surges, prices collapse. These unofficial flows complicate forecasting and increase the difficulty of regulating markets.

Speculative stockpiling is another factor shaping price movements. Chinese firms, particularly refiners and battery manufacturers, often purchase large volumes of lithium, cobalt, nickel, or rare earths when prices are low, filling warehouses in anticipation of future demand. This can create artificial scarcity in the short term, driving prices upward even when supply is adequate. Conversely, when companies sell stockpiled material, prices may fall abruptly. Market participants must therefore interpret not only supply and demand but perceived intent—an inherently speculative exercise.

Price reporting agencies like Argus and Asian Metal play crucial roles in markets where no exchanges exist. Their published prices influence contracts, investor sentiment, and procurement decisions. But because these assessments rely on private data from transactions that may be opaque or unrepresentative, there is room for error or manipulation. Companies often hedge by relying on multiple sources or negotiating formula-based contracts tied to a basket of indices.

The emergence of long-term offtake agreements reflects buyer efforts to escape volatility. Automakers, battery manufacturers, and tech companies increasingly sign multi-year deals with lithium, nickel, and cobalt producers, often including price bands or indexed formulas. These agreements create stability but can backfire when spot prices fall below contract terms, prompting tension or renegotiation. Yet without such contracts, many mining projects cannot secure financing. Long-term deals thus sit at the intersection of financial risk and industrial necessity.

Financialization affects not only prices but also geopolitics. When market speculation exaggerates scarcity, governments respond with stockpiles, export controls, or investment incentives. Speculative bubbles can distort policymaking, prompting ill-timed regulations or encouraging unviable projects that collapse during downturns. Conversely, price collapses may lead governments to underestimate long-term need, delaying critical investments until supply once again tightens.

The traders themselves act as geopolitical intermediaries. Trafigura, Glencore, and others have financed infrastructure in resource-rich countries, supported state enterprises, or managed logistics during political instability. Their networks often allow them to operate in regions where governments or corporations cannot. This gives traders a quasi-diplomatic role: they manage cross-border risk, maintain relationships with unstable jurisdictions, and keep material flowing even during crises.

Meanwhile, financial innovations continue to reshape markets. Futures contracts for lithium hydroxide, introduced by the LME in 2021, have grown slowly but signal an effort to reduce opacity. Options, exchange-traded products, and physical-backed funds are emerging around battery metals. These instruments can improve transparency but also introduce new layers of speculation. In thin markets, even modest financial flows can distort prices.

As demand for critical minerals continues to grow—driven by EV adoption, renewable energy, semiconductors, and digitization—price volatility will remain a central challenge. Managing it will require coordinated efforts across the value chain: stable policy, diversified supply, improved transparency, more robust financial instruments, and better inventory management. Capital allocation is increasingly decisive in shaping mineral markets. Investment flows, sovereign wealth strategies, and financial speculation determine which projects advance, which stall, and how supply chains consolidate. In an environment of scarcity and competition, finance becomes a structural force in material power.

CHAPTER 40 – INVESTMENTS, SOVEREIGN WEALTH FUNDS, AND MARKET SPECULATION

The rise of critical minerals as the backbone of twenty-first-century technologies has triggered a profound shift in global investment patterns. What were once niche extractive industries have become central assets in the portfolios of private equity firms, sovereign wealth funds, pension funds, multilateral banks, corporate investors, and state-backed development agencies. Critical minerals are no longer just commodities; they are strategic financial instruments linked to energy security, technological competitiveness, and national power. As a result, investment flows into the sector have surged, becoming increasingly intertwined with geopolitics and speculation.

The scale and composition of investments in critical minerals have changed dramatically over the past decade. Historically, mining was dominated by specialized resource funds and major industry players. Today, investment flows come from a far wider array of sources: automotive corporations seeking battery metal security, tech giants protecting supply chains for electronics and data centers, sovereign wealth funds looking for long-term returns, and states seeking strategic leverage. This diversification of investors reflects the recognition that minerals are foundational to economic growth and political stability.

At the center of this transformation are sovereign wealth funds (SWFs). These state-owned investment vehicles—managing trillions of dollars globally—have emerged as major actors in the critical minerals race. Their financial power, long-term investment horizon, and political backing allow them to shape global supply chains in ways private investors cannot. Middle Eastern SWFs, particularly those in Saudi Arabia, the United Arab Emirates, and Qatar, are using their immense capital reserves to diversify beyond hydrocarbons by investing directly in mining, battery materials, and refining. Saudi Arabia's Public Investment Fund has backed copper, nickel, and lithium projects through Ma'aden and international joint ventures. Abu Dhabi's Mubadala has invested in energy-transition metals, advanced materials, and battery supply chains. These funds often act not only for financial returns but also to position their countries within the emerging industrial order.

Asian SWFs also play transformative roles. Singapore's Temasek and GIC have backed battery recycling, materials science startups, and renewable infrastructure. South Korea's KIC and Japan's GPIF have invested in mining projects indirectly through commodity funds and infrastructure bonds. China's sovereign investment mechanisms—including the China Investment Corporation and state policy banks—channel capital into overseas mining assets, securing long-term supply for Chinese industry. These funds are deeply embedded in China's broader resource strategy, supporting acquisitions in Africa, Latin America, and Australia.

Western SWFs, such as Norway's Government Pension Fund Global or Canada's various provincial funds, tend to avoid direct mining exposure due to sustainability concerns, but they influence markets through thematic investments, infrastructure financing, and support for "green metal" initiatives. They may not buy mines, but they back renewable technologies that increase demand for minerals, shaping long-term market expectations.

Investment decisions by SWFs ripple across financial markets. When a sovereign fund signals interest in a particular metal—lithium, copper, nickel, or rare earths—it can trigger a wave of private investment and speculative enthusiasm. Conversely, if a fund exits a sector due to ESG concerns, valuations may collapse. This reflexive relationship between capital flows and market sentiment contributes to the cyclical volatility explored in the previous chapter.

Private investors play an equally important role. Private equity and venture capital firms have poured billions into battery materials, recycling technologies, solid-state battery startups, mining exploration, and processing technologies. Some funds specialize in early-stage mining projects that traditional banks deem too risky. Others invest in advanced materials research, supporting the development of alternatives to scarce minerals. The speculative nature of these investments reflects not just hope for high returns but a belief that governments will continue to accelerate the energy transition, ensuring long-term demand.

Institutional investors—pension funds, insurance companies, endowments—have been slower to engage directly due to risk profiles and ESG concerns. However, as critical minerals become recognized as essential for decarbonization, institutional capital is gradually shifting. Some funds are revising their guidelines to allow investments in "responsibly certified" mining operations that meet strict environmental and social standards. Institutional capital could become a stabilizing force, providing long-term financing to projects that currently rely on higher-risk private equity.

Corporate investments are reshaping the landscape even more dramatically. Automakers, battery manufacturers, and technology firms are increasingly financing mines, refineries, and chemical plants. These investments—discussed in depth in earlier chapters—represent a form of vertical integration that extends far upstream. Tesla, Ford, GM, Volkswagen, and Toyota have all secured long-term mineral supply through direct investments, joint ventures, or convertible financing agreements. Apple and Samsung have quietly supported cobalt and rare earth supply chain reforms. These corporate investments reduce reliance on traders and insulate companies from market volatility, but they also contribute to supply chain fragmentation as industries create exclusive partnerships and closed loops.

Traditional banks and multilateral development institutions still play vital roles, especially in politically sensitive regions. The World Bank, IMF, African Development Bank, and Asian Development Bank fund infrastructure, environmental oversight, and capacity-building programs in mining-heavy countries. They may not invest directly in mines, but their projects—roads, power plants, logistics hubs—enable mining development. Their guidelines help shape governance standards, influencing how national governments manage mineral wealth. Nonetheless, these institutions often move slowly, leaving space for faster-moving state-backed funds from China, the Gulf, or emerging economies.

Speculation—both informed and opportunistic—adds another layer of complexity. Commodity hedge funds increasingly trade lithium, cobalt, nickel, and rare earths, even when liquid futures markets do not exist. They rely on price-reporting agencies, bilateral deals, and long-term macro projections. Physical-backed investment products—similar to gold ETFs—have begun appearing for cobalt, nickel, and lithium, inviting retail and institutional speculation. These vehicles create new sources of demand unrelated to industrial consumption, amplifying price cycles.

Market expectations shape investment decisions. If investors believe that EV adoption will accelerate beyond forecasts, they may flood lithium or nickel markets with capital, creating a price bubble. If they fear substitution—such as sodium-ion batteries reducing demand for lithium—they may shift capital away prematurely. These sentiment-driven swings can leave mining companies underfunded in critical moments or encourage overinvestment in unneeded capacity. Financial markets thus influence physical supply, not merely reflect it.

National security concerns further complicate investment flows. Governments increasingly screen foreign investments in critical minerals, blocking acquisitions that might give rival states undue influence. The United States, Europe, Australia, and Canada have all strengthened foreign investment review mechanisms, especially regarding Chinese acquisitions. This geopolitical filter reduces the pool of potential investors for some projects and increases competition for those deemed strategically aligned. It may also slow project development when government scrutiny delays investment timelines.

One of the most important trends in recent years is the rise of strategic partnerships between states, corporations, and financial institutions. Governments provide loan guarantees or equity stakes to reduce risk for private investors. Corporations commit to long-term purchase agreements that secure project revenues. Banks offer project financing backed by sovereign guarantees. This hybrid investment model is beginning to dominate the sector, reflecting the reality that no single actor—company, investor, or state—can secure the mineral supply chain alone.

Speculation, investment, and geopolitical strategy are curling around one another in ways that make critical mineral markets uniquely volatile and politically sensitive. The materials that power the energy transition are not simply bought and sold; they are sought after, fought over, financed in complex ways, and embedded in long-term bets about the future of global industry.

As sovereign wealth funds, private investors, corporations, and governments deepen their involvement, the power balance within the global mineral system will continue to evolve. Environmental, social, and governance standards are no longer peripheral to mineral markets. Ethical scrutiny and sustainability frameworks increasingly shape where capital flows and which supply chains are considered viable. In a high-stakes competitive environment, responsibility itself becomes a strategic constraint.

CHAPTER 41 – ESG, ETHICS, AND RESPONSIBLE SUPPLY CHAINS

As the global demand for critical minerals accelerates, the pressure on companies, investors, and governments to ensure ethical, transparent, and environmentally sustainable supply chains has intensified. ESG—environmental, social, and governance—has become more than a corporate label or marketing tool. It is now a central pillar of industrial legitimacy, regulatory compliance, and geopolitical strategy. The minerals driving batteries, renewable energy, semiconductors, and defense technologies are extracted and processed through complex, often opaque global systems. These systems carry risks: environmental degradation, labor exploitation, corruption, community displacement, hazardous waste, and geopolitical entanglement. ESG frameworks attempt to address these risks, but the gap between ambition and reality remains wide.

The rise of ESG in critical minerals is driven by three converging forces. First, public and political scrutiny has reached unprecedented levels. Consumers are increasingly aware that their electric vehicles, smartphones, and solar panels rely on minerals extracted from vulnerable regions. Media coverage of child labor in cobalt mining, toxic tailings spills, or illegal rare earth operations has generated global concern. Second, governments and regulators are imposing stricter standards on supply chain transparency, forcing companies to disclose sourcing practices and due diligence processes. Third, investors—particularly institutional funds—are integrating ESG criteria into their decisions, refusing to finance projects that expose them to reputational or regulatory risk.

Environmental concerns are the most widely recognized dimension of ESG. Mining and processing of critical minerals can cause significant ecological damage if mismanaged. Lithium brine extraction in arid regions consumes large quantities of water, creating tensions with local communities. Nickel extraction through high-pressure acid leaching generates substantial waste and carbon emissions. Rare earth separation produces radioactive byproducts that require strict containment. Graphite purification often uses hydrofluoric acid, creating hazardous waste streams. These impacts challenge the narrative that critical minerals are inherently "green." They are essential to decarbonization, but their extraction and processing can create serious environmental burdens.

Companies are responding by investing in cleaner technologies, adopting stricter environmental standards, and collaborating with governments to improve regulatory frameworks. Some mining operations are transitioning to renewable energy, electrifying equipment, or implementing closed-loop water systems. Chemical processors are upgrading filtration and waste treatment technologies. Automakers are demanding low-carbon materials for their EV batteries, creating incentives for greener production methods. Yet these improvements are unevenly distributed. Many operations in emerging economies lack the infrastructure or regulatory enforcement needed to meet high environmental standards, leaving global supply vulnerable to environmental malpractice.

Social risks present an equally complex challenge. The Democratic Republic of Congo—the world's primary source of cobalt—faces persistent issues with artisanal mining, including unsafe working conditions, informal labor, and child labor. Although artisanal mining provides livelihoods for hundreds of thousands of people, it also exposes workers to toxic conditions and exploitation. Attempts by international companies to eliminate artisanal cobalt from supply chains often backfire, driving miners further into precarious conditions and reducing oversight.

Responsible sourcing initiatives seek to integrate artisanal miners into formal supply chains rather than exclude them. These programs provide training, safety equipment, fair compensation, and monitoring systems. Some companies partner with NGOs to implement on-site controls, trace materials through blockchain systems, or create cooperative structures that empower local miners. While progress is slow, these approaches acknowledge a fundamental truth: sustainable development requires engaging with informal economies, not erasing them.

Other social issues include land rights, consultation with Indigenous peoples, compensation for displaced communities, and the social impacts of large-scale industrial operations. In Latin America, Indigenous groups have resisted lithium and copper projects that threaten their water sources and cultural heritage. In Africa, mining concessions sometimes overlap with community land, leading to disputes and mistrust. ESG frameworks increasingly require companies to obtain free, prior, and informed consent (FPIC) from affected populations, but implementation remains inconsistent.

Governance risks—corruption, bribery, transparency failures, and weak institutions—can undermine even the most well-intentioned ESG programs. Some resource-rich countries suffer from unstable regulatory environments, political interference, or elite capture of mining revenues. This creates uncertainty for investors and fuels local resentment. International initiatives such as the Extractive Industries Transparency Initiative (EITI) aim to improve governance by requiring disclosure of payments, contracts, and ownership structures. Though influential, these efforts face challenges in regions where governance remains fragile.

In recent years, major consuming regions have enacted regulations that push companies toward responsible sourcing. The European Union's Battery Regulation sets strict due diligence requirements for battery materials, including cobalt, lithium, nickel, and graphite. Companies must verify that these minerals are not associated with environmental harm, human rights abuses, or corruption. The EU's Corporate Sustainability Due Diligence Directive (CSDDD) further obligates large companies to monitor and address risks across their entire supply chain, not only direct suppliers.

In the United States, the Uyghur Forced Labor Prevention Act (UFLPA) prohibits the import of goods connected to forced labor in Xinjiang. This has major implications for polysilicon, lithium, and other materials processed in the region. Companies must provide documentation proving that materials are sourced from compliant suppliers—a challenging requirement given the opacity of certain supply chains. U.S. agencies have also issued guidance for "responsible sourcing of critical minerals," aligning national security strategy with ESG considerations.

Corporate responses to ESG vary widely. Some companies embrace ESG as a competitive advantage. Automakers use responsible sourcing as a marketing tool, highlighting sustainable battery materials or ethical labor practices. Tech companies publish detailed supply chain maps and commit to 100 percent responsible sourcing.

Mining companies seek certifications from organizations such as the Initiative for Responsible Mining Assurance (IRMA), highlighting their adherence to rigorous environmental and social standards. These certifications can improve access to capital and attract investors concerned about reputational risk.

Yet ESG implementation remains inconsistent. Some companies apply high standards only in jurisdictions with strong regulatory oversight, while overlooking risks in more vulnerable regions. Others produce glossy sustainability reports that mask limited action. The disconnect between corporate narrative and operational reality has led to criticism that ESG is sometimes used to deflect scrutiny rather than promote genuine reform. Still, ESG pressures are increasingly difficult to ignore. Investors, regulators, and civil society groups are demanding evidence of real impact, not aspirational promises.

A growing trend is the integration of traceability technologies. Blockchain platforms, digital ledgers, and advanced QR coding systems allow companies to track materials from mine to battery. These tools help prevent fraud, improve transparency, and verify ESG compliance. They also support the development of "digital passports" for batteries, an EU initiative requiring detailed disclosure of material origins, carbon footprint, and recycling content. These passports could become global norms, allowing consumers to understand the lifecycle of the materials powering their vehicles and devices.

Recycling plays a central role in responsible supply chains. By recovering lithium, nickel, cobalt, and other materials from spent batteries, recycling reduces pressure on mining, lowers environmental impact, and supports circular economies. Many companies are investing in recycling technologies as part of their ESG strategies, recognizing that end-of-life management is as important as responsible extraction. Governments incentivize recycling through subsidies, mandates, and research funding, helping establish closed-loop systems that reduce dependence on virgin materials.

ESG and responsible supply chains are not optional—they are fundamental to the future of the critical minerals economy. Companies that fail to adopt high standards risk losing access to capital, markets, and consumer trust. Governments that ignore ESG concerns risk social unrest, environmental degradation, and international isolation. Investors that neglect ESG risks face stranded assets and financial losses.

The next and final chapter of Part V will examine how recycling firms and new materials startups are reshaping the future of the sector—offering pathways to reduce ESG risks, diversify supply, and accelerate the transition to sustainable industrial ecosystems.

CHAPTER 42 – THE RISE OF RECYCLING FIRMS AND NEW MATERIALS STARTUPS

As the world races to secure the minerals that power its technologies, the focus is often placed on mines, refineries, and geopolitics. But a quieter revolution is emerging—one that may ultimately prove just as consequential for global supply, sustainability, and industrial resilience. Recycling firms and new materials startups are transforming how economies source critical minerals, how companies manage technological risk, and how nations reduce dependence on volatile global supply chains. These emerging actors represent the most innovative and fast-evolving segment of the critical minerals ecosystem, offering pathways that complement traditional extraction but also challenge its dominance.

The rise of recycling is driven by three structural forces. First, the demand for lithium, nickel, cobalt, and rare earths is growing exponentially as electric vehicles, grid storage, consumer electronics, and renewable systems proliferate. Second, mining and refining capacity cannot expand fast enough to meet projected needs, especially given long permitting timelines and geopolitical bottlenecks. Third, ESG pressures push companies and governments to reduce environmental impact, manage waste responsibly, and move toward circular economies. Recycling—once viewed as marginal—has thus become a strategic imperative.

Modern battery recycling firms operate at the frontier of materials science. They employ hydrometallurgical and pyrometallurgical techniques to extract high-purity lithium, nickel, cobalt, manganese, and graphite from spent batteries and manufacturing scrap. Unlike traditional recycling, which focuses on recovering metals from mixed waste streams, battery recycling produces materials that can re-enter battery supply chains with purity levels comparable to primary production. In many cases, recycled materials have a lower carbon footprint and more predictable composition than mined equivalents.

A growing number of companies have emerged to lead this transformation. Firms in North America, Europe, and Asia are building large-scale recycling facilities capable of processing tens of thousands of tons of battery material annually. Their rise reflects growing corporate demand for secure, low-carbon, traceable materials. Automakers partner with recyclers as part of their vertical integration strategies, securing closed-loop supply agreements that reduce exposure to volatile commodity markets. Battery manufacturers integrate recycling into their operations, using scrap recovery to improve material efficiency and reduce costs.

One of the most striking features of the recycling sector is its technological diversity. Some firms specialize in black mass processing—the intermediate material created by shredding spent batteries. Others focus on refining specific metals, producing battery-grade lithium carbonate or nickel sulphate. Some companies innovate in mechanical separation, recovering copper foil, aluminum casings, plastics, and other components.

A few specialize in rare earth magnet recycling, extracting neodymium, praseodymium, and dysprosium from end-of-life motors and generators. This diversity mirrors the complexity of the supply chains they serve.

Recycling also carries strategic implications for geopolitics. Countries with large automotive industries—such as the United States, Germany, Japan, and South Korea—see battery recycling as a way to build domestic material security. Recycling reduces dependence on imports from politically sensitive regions and supports local industrial ecosystems. Europe, in particular, views recycling as essential to its strategy of achieving autonomy in battery materials. The EU's Battery Regulation mandates minimum recycled content for lithium, cobalt, and nickel by the end of the decade, ensuring that recycling becomes an integrated part of the industrial landscape.

Yet recycling alone cannot meet near-term demand. The volume of end-of-life batteries is still relatively small, and the bulk of material processed today comes from manufacturing scrap. It will take years before significant quantities of EV batteries reach end of life. Still, recycling firms are building capacity in anticipation of future flows, creating a foundation for long-term circularity.

Parallel to the recycling revolution is the rise of new materials startups—companies that seek to reduce or replace critical minerals through scientific innovation. These firms develop alternative chemistries, new battery architectures, rare-earth-free magnets, novel semiconductor materials, and breakthrough manufacturing processes designed to mitigate supply risks.

In the battery sector, startups are pioneering materials that use less lithium, nickel, or cobalt—or avoid them entirely. Sodium-ion batteries are one of the most promising alternatives. They use abundant sodium instead of lithium and do not require nickel or cobalt. Although they have lower energy density, they offer advantages in cost, safety, and scalability. Several companies in China and Europe have begun producing sodium-ion cells at commercial scale, targeting grid storage and low-cost vehicle segments. Their progress suggests that sodium-ion could become an important complement to lithium-ion technologies.

Silicon-rich anodes represent another breakthrough. By incorporating high percentages of silicon into anodes, companies aim to dramatically increase energy density and charging speed. These materials reduce dependence on graphite, a supply chain dominated by China. While silicon anodes pose challenges such as expansion during cycling, startups are developing coatings and composite structures to mitigate degradation. If successful, silicon-based anodes could reshape demand patterns for critical minerals.

In the realm of cathode materials, startups are exploring manganese-rich chemistries, cobalt-free formulations, and advanced high-nickel designs that balance supply risk with performance. Some firms aim to produce lithium iron phosphate (LFP) cathodes outside China, supporting diversification. Others pursue solid-state batteries that use entirely different material systems and promise significant advances in safety and performance. These technologies, while still developing, illustrate how innovation can alter mineral demand trajectories.

Rare-earth-free magnet technologies are another frontier. Startups are developing magnets that use iron-nitride, ferrites, or other compositions to reduce or eliminate neodymium, praseodymium, and dysprosium. While performance gaps remain for high-temperature and high-power applications, progress is accelerating. If these materials reach commercial maturity, they could reduce dependence on one of the world's most politically concentrated supply chains.

In semiconductors, startups are exploring new materials such as gallium oxide, silicon carbide, and perovskites for power electronics and solar applications. These materials promise higher efficiency and performance, but their supply chains are still emerging. Innovation in semiconductor materials requires close coordination between research institutions, manufacturers, and government agencies—reflecting the high-stakes nature of the field.

What distinguishes new materials startups from traditional industry players is their speed, flexibility, and willingness to challenge established assumptions. They operate in a space of scientific uncertainty, where breakthroughs carry immense potential rewards but also significant risk. Investors are increasingly funding these ventures, recognizing that technological substitution is one of the most powerful tools for mitigating supply chain vulnerability.

Yet innovation cannot fully displace the need for responsibly sourced minerals. Most emerging technologies still require some combination of lithium, nickel, cobalt, copper, graphite, or rare earths. Recycling and substitution can reduce demand growth, smooth volatility, and enhance resilience, but they cannot eliminate mining. The world will need more minerals in the decades ahead—but it will also need smarter ways of sourcing them.

Recycling firms and new materials startups complement each other. Recycling addresses short-term supply constraints, reduces environmental impact, and supports circularity. New materials offer long-term pathways to diversify away from fragile supply chains and reduce geopolitical dependencies. Together, they form the innovation ecosystem at the end of the supply chain—an ecosystem that is increasingly critical for industrial competitiveness and strategic autonomy.

The rise of these companies signals a broader transformation. The future of critical minerals will not be defined solely by extraction, refining, or geopolitics. It will also be shaped by innovation—by the ability of scientists, engineers, and entrepreneurs to rethink the materials that power the modern world. In the chapters that follow, this book will turn toward long-term scenarios, strategic planning frameworks, and pathways for building a resilient, sustainable, and equitable mineral future.

PART VI – SUSTAINABILITY, INNOVATION, AND FUTURE SCENARIOS

CHAPTER 43 – ENVIRONMENTAL AND SOCIAL IMPACTS OF MINING

The accelerating global demand for critical minerals has brought mining to the center of the world's attention. Once perceived as a distant, industrial activity happening far from everyday life, mining has reemerged as a visible, political, and environmental issue shaping debates about climate change, justice, and technological progress. The minerals required for electric vehicles, renewable energy systems, consumer electronics, and advanced manufacturing must come from somewhere—and that "somewhere" is often a fragile landscape inhabited by communities with limited political power. As the clean-energy transition expands, so too does the tension between the need for more minerals and the responsibility to obtain them in ways that do not replicate the harms of the past. Understanding the environmental and social impacts of mining is therefore essential for any realistic conversation about the future of critical materials.

The environmental footprint of mining is considerable. Extracting minerals disrupts ecosystems, alters landscapes, consumes water, and generates waste. These impacts vary by mineral type, geology, and technology, but several common patterns emerge across the industry. For open-pit mining, the most widespread technique, vast areas of land must be excavated, removing vegetation and soil. This disrupts habitats, affects biodiversity, and creates long-lasting scars on the land. In arid regions, water-intensive operations strain local supplies and increase competition between industrial and community needs. In tropical regions, mining can accelerate deforestation and contribute to erosion and sedimentation in rivers. Even in well-regulated jurisdictions, preventing contamination of soil and water remains a major challenge.

Tailings—the waste slurry left after minerals are separated from ore—pose one of the most significant environmental risks. Tailings dams, some of which contain millions of cubic meters of slurry, have a history of catastrophic failures. Several high-profile dam collapses have caused severe environmental damage, killing communities downstream and contaminating river systems for generations. These incidents have fueled global concern and prompted calls for stronger oversight. Yet many tailings facilities around the world remain underregulated, poorly monitored, or located in regions prone to extreme weather. As mining expands, tailings management will become an increasingly central part of environmental risk.

Critical minerals often involve chemically intensive processing methods. Rare earth separation, for example, requires large quantities of acids and produces radioactive residues. When disposal is mismanaged, these residues can leak into groundwater or create long-term contamination zones. Lithium extraction from brine fields, especially in South America's "Lithium Triangle," relies on evaporative ponds that concentrate minerals but also lead to substantial water loss. This extraction method has sparked debate over its impact on Indigenous communities, wetlands, and fragile ecosystems. While scientific studies disagree on the magnitude of the effect, the perception of harm is real and affects social legitimacy.

Nickel extraction—particularly through high-pressure acid leaching—generates large volumes of waste and emits significant greenhouse gases. In Indonesia and the Philippines, where nickel mining has expanded rapidly, environmental concerns include deforestation, river contamination, and destruction of coastal ecosystems. Even when mining companies comply with national regulations, local communities often experience the cumulative effects of multiple mines operating in close proximity. Meanwhile, graphite purification in parts of China has contributed to air pollution and hazardous waste, prompting protests and government crackdowns. These examples illustrate a critical reality: even minerals essential for clean technologies can be produced in ways that are anything but clean.

Social impacts are equally significant. Mining often occurs in regions where governance is weak, institutions are fragile, and communities have limited influence over political decisions. This combination increases the risk of human rights violations, corruption, and exploitation. Artisanal and small-scale mining—especially in the Democratic Republic of Congo for cobalt—continues to raise serious concerns about child labor, unsafe working conditions, and lack of formal oversight. Artisanal mining provides essential income for local populations, yet it exposes miners to dangerous tunnels, toxic dust, and unstable geology. Many efforts to improve safety focus on integrating artisanal miners into formal supply chains, providing training, equipment, and stable pricing. But these programs face political, cultural, and logistical challenges.

Community displacement is another major social impact. Large mining projects often require land acquisition on a scale that disrupts entire villages or traditional territories. In Latin America, Indigenous communities have resisted lithium and copper projects that threaten water rights, sacred sites, or local livelihoods. In Africa, mining concessions sometimes overlap with farming land, grazing areas, or forests critical for community survival. When consultation processes are weak or manipulated, mistrust deepens and conflict becomes likely. Even when compensation is offered, it rarely reflects the full cultural, economic, and psychological value of what is lost.

Social impacts extend beyond displacement. Mining can bring economic opportunities but also social inequality, inflation, and strain on local infrastructure. The influx of workers can alter community dynamics, creating tensions or increasing pressure on housing, healthcare, and services. Mining revenues, when poorly managed, may be captured by elites rather than benefiting the broader population. Corruption in resource-rich regions is a persistent challenge, undermining governance and reducing the likelihood that local communities will see tangible benefits.

Health impacts also feature prominently. Dust from mining operations can contribute to respiratory illnesses. Contaminated water supplies pose risks of heavy metal poisoning. Noise pollution and traffic from mining-related transport can affect daily life. While modern mining standards require mitigation measures, enforcement varies dramatically. In countries with limited regulatory capacity, environmental and social protections may exist more on paper than in practice.

Yet mining does not always follow a negative path. There are examples of responsible mining that minimize impacts, engage communities meaningfully, and generate long-term benefits. In some jurisdictions—such as Canada, Australia, and Scandinavia—strong regulatory frameworks, community participation, and transparent governance help ensure higher standards. Modern practices emphasize rehabilitation, biodiversity offsets, tailings safety, water recycling, and carbon reduction. Some mines operate with renewable energy or electrified equipment to reduce emissions. Others implement advanced monitoring systems that provide real-time data on environmental performance. These examples show that when regulation, technology, and community engagement align, mining can contribute to sustainable development.

The challenge lies in extending these best practices globally. Regions with the richest mineral deposits often lack the governance structures that support responsible mining. This creates a paradox: those who most depend on mining for development are also those most vulnerable to its harms. International frameworks aim to fill some of these gaps. Certifications, due diligence standards, global industry initiatives, and NGO-led assessments help create incentives for compliance. However, these mechanisms often lack enforcement power and may be resisted by countries that see them as barriers to development or as forms of neocolonial oversight.

As the world ramps up mineral extraction to power the energy transition, these environmental and social impacts become global concerns. Achieving climate goals will require unprecedented quantities of minerals, but pursuing them without regard for ecosystems and communities would undermine the very goals the transition seeks to achieve. This tension—between sustainability aspirations and industrial imperatives—defines the debate ahead.

New extraction frontiers are pushing mineral supply beyond conventional boundaries. Deep-sea mining and biomining offer technological promise while introducing profound uncertainty, risk, and ethical tension. How these approaches are governed will influence not only supply security, but the legitimacy of future resource strategies.

CHAPTER 44 – DEEP-SEA MINING, BIOMINING, AND NEW EXTRACTION FRONTIERS

As global demand for critical minerals accelerates, the limits of conventional mining are becoming increasingly apparent. Terrestrial deposits are often concentrated in politically sensitive regions, environmentally fragile ecosystems, or jurisdictions with complex social challenges. At the same time, tightening regulations, lengthy permitting processes, community resistance, and rising ESG expectations make new land-based mining projects difficult to advance. In this context, governments, companies, and research institutions are exploring unconventional extraction frontiers: deep-sea mining, biomining, enhanced geothermal recovery, brines from oilfields, and other emerging techniques that challenge traditional assumptions about where minerals can come from. These frontiers may reshape the geography of resource extraction—but they also raise profound environmental, ethical, and governance questions.

Among these new domains, deep-sea mining has received the most attention and provoked the most intense controversy. The seabed hosts vast quantities of minerals essential for batteries, renewable energy technologies, and advanced electronics. Polymetallic nodules on abyssal plains contain high concentrations of nickel, cobalt, manganese, and rare earth elements. Seafloor massive sulfides at hydrothermal vents contain copper, zinc, silver, and gold. Cobalt-rich crusts on seamounts offer another potential resource. These deposits appear attractive for several reasons: they are abundant, they occur in predictable environments, and they contain metals in concentrations often higher than land-based ores.

Supporters of deep-sea mining argue that it could alleviate supply shortages, reduce reliance on terrestrial regions with complex governance challenges, and reduce the environmental footprint associated with deforestation, water extraction, and toxic waste from land-based mines. They also highlight that deep-sea nodules require no blasting, no tailings dams, and no removal of vegetation. Some proponents suggest that deep-sea mining could provide the minerals needed for the energy transition without reproducing the terrestrial impacts discussed in the previous chapter.

Yet these claims face serious scrutiny. Deep-sea ecosystems remain poorly understood, hosting unique species adapted to extreme conditions—many of which have not yet been discovered. Extraction techniques involve harvesting nodules with large underwater vehicles, creating sediment plumes, noise, and light disturbances that can affect biodiversity over wide areas. The deep ocean is a fragile environment that recovers at geological timescales, not human ones. Disturbances may be irreversible. Scientists warn that sediment plumes could travel long distances, smothering organisms and altering biogeochemical cycles. The extraction of nodules also removes habitats that took millions of years to form. The lack of long-term ecological data intensifies concerns, making deep-sea mining one of the most polarizing environmental debates of the decade.

The governance framework for deep-sea mining is also contested. Much of the seabed lies beyond national jurisdictions, overseen by the International Seabed Authority (ISA). The ISA is tasked with balancing resource development with environmental protection, but critics argue it lacks the scientific data, enforcement mechanisms, and transparency needed to regulate such an immense responsibility. Several nations and corporations have applied for exploratory licenses, while others—including scientists, NGOs, and entire countries—are calling for moratoria or bans until adequate scientific understanding and regulatory safeguards are established. This conflict illustrates the dilemma of new extraction frontiers: technological enthusiasm colliding with ecological uncertainty.

Another promising frontier is biomining, which uses microorganisms to extract metals from ores, waste, or low-grade deposits. Unlike conventional methods that require high heat, pressure, or toxic chemicals, biomining employs bacteria that naturally oxidize metal-bearing minerals. This process can extract copper, gold, nickel, cobalt, and rare earth elements with lower energy requirements and reduced environmental impact. Biomining is already used commercially in copper extraction in countries like Chile, and research continues to expand its potential applications.

The appeal of biomining lies in its scalability, environmental benefits, and ability to process materials unsuitable for traditional techniques. Tailings, waste rock, or extremely low-grade ores can become resource streams rather than liabilities. Biomining also offers potential for in situ extraction, where microbes are introduced into rock formations to dissolve metals underground, eliminating the need for open-pit mines. This could transform how society approaches legacy mining sites and closed deposits, reducing environmental footprints while extending the life of known resources.

Rare earth extraction presents another compelling opportunity for biomining. Certain bacteria can selectively bind or dissolve rare earth elements, enabling more environmentally friendly separation processes. If scaled successfully, bio-based rare earth extraction could reduce dependence on chemically intensive solvent extraction—the dominant method in global supply chains. Although still in development, these biological pathways offer a promising direction for reducing environmental harm in one of the world's dirtiest segments of mineral processing.

Enhanced geothermal systems (EGS) represent another unconventional source of critical materials. Geothermal brines often contain dissolved lithium, manganese, zinc, and other metals. Technologies for extracting minerals from geothermal fluids are advancing rapidly, especially in regions like the United States, Germany, and Iceland. If successful at scale, geothermal mineral extraction could produce lithium with dramatically lower emissions and water use compared to traditional methods. The dual benefit—renewable energy production and mineral recovery—makes EGS an attractive option for nations seeking energy and resource security simultaneously.

Oilfield brines also offer surprising potential. As part of routine oil extraction, large volumes of mineral-rich water are brought to the surface. These fluids often contain lithium, bromine, strontium, and other elements. Companies are now developing technologies to extract these minerals cost-effectively before reinjecting the water underground. This could repurpose existing infrastructure, reduce environmental impacts compared to new mining projects, and expand supply without disturbing new ecosystems. The approach aligns with broader efforts to reduce the carbon footprint of oil and gas operations while supporting clean-energy supply chains.

Urban mining represents yet another new frontier. As discussed in earlier chapters, the materials embedded in electronic waste—circuit boards, magnets, batteries, cables—constitute a vast above-ground mine. New startups employ advanced hydrometallurgical and pyrometallurgical techniques, as well as AI-driven sorting systems, to recover metals from discarded devices. This approach reduces pressure on natural ecosystems and transforms waste into a valuable resource. While urban mining is not new, its scale and sophistication are poised to increase dramatically in the coming decades, especially as governments implement stricter recycling mandates and digital product passports requiring manufacturers to disclose end-of-life pathways.

Each of these new extraction frontiers presents its own set of trade-offs. Deep-sea mining may provide abundant resources but risks damaging ecosystems that humanity barely understands. Biomining offers environmental advantages but faces technical limitations, long processing times, and uncertain scalability. Geothermal and oilfield brines reduce land disturbance but depend on specific geological conditions and high capital investment. Urban mining reduces the need for primary extraction but cannot yet meet full demand for rapidly growing sectors like electric vehicles.

The central challenge is not choosing a single solution but developing a balanced portfolio of extraction technologies. The energy transition requires enormous quantities of minerals, and relying exclusively on traditional mining will exacerbate environmental and social tensions. Diversifying supply through emerging techniques can reduce geopolitical vulnerabilities, lower ecological impacts, and support long-term sustainability goals. Yet none of these approaches is free from risk. Scientific rigor, transparent governance, and robust oversight are essential to ensure that new extraction frontiers do not repeat the mistakes of the past.

Decisions over new extraction pathways extend far beyond short-term supply calculations. Deep-sea mining, biomining, and geothermal extraction are political choices shaped by regulation, public trust, scientific uncertainty, and corporate incentives. How societies weigh these factors will determine whether material security deepens dependence on extraction or accelerates a shift toward fundamentally different resource models.

CHAPTER 45 – CIRCULAR ECONOMY AND CLOSED-LOOP SYSTEMS

As global demand for critical minerals accelerates, the world is confronting a paradox at the heart of the clean-energy transition: the technologies designed to decarbonize modern economies depend on materials sourced through carbon-intensive, environmentally disruptive, and geopolitically fraught supply chains. Lithium-ion batteries, permanent magnets, photovoltaic cells, semiconductors, and high-performance alloys all require minerals extracted through processes that often carry social and ecological costs. This tension has pushed policymakers, corporations, and researchers toward a single conclusion: the future of mineral supply cannot rely solely on extraction. It must increasingly embrace circular economy principles and closed-loop systems—approaches that minimize waste, extend product lifecycles, and recover valuable materials from end-of-life products.

The circular economy represents a fundamental shift away from the traditional linear model of "take, make, dispose." Instead, it envisions a system in which products are designed to be reused, repaired, disassembled, and recycled, with materials re-entering industrial cycles in continuous loops. For critical minerals, this shift promises three major benefits: reducing pressure on mining, lowering environmental impacts, and increasing supply security. But building a circular mineral economy is far more complex than recycling plastic bottles or scrap steel. It requires technological innovation, regulatory reform, industrial redesign, and new business models that challenge entrenched assumptions about how materials are used.

Batteries are at the center of circular economy discussions. Lithium-ion batteries contain valuable metals such as lithium, nickel, cobalt, manganese, and copper. As electric vehicles proliferate, the volumes of spent batteries are expected to grow dramatically by the end of the decade. These batteries represent a vast above-ground resource—often called "urban ore." Unlike primary ores buried deep underground, battery materials are already concentrated and chemically accessible. Recovering them is not merely an environmental imperative; it is economically strategic. Closed-loop battery systems can significantly reduce dependence on geographically concentrated mines and refined materials dominated by a handful of countries.

Leading battery manufacturers and automotive companies are moving rapidly to integrate recycling into their supply chains. This integration is driven partly by ESG considerations but also by the need to stabilize costs and secure long-term access to materials. Some companies have adopted designs that facilitate disassembly and material recovery. Others have partnered with recycling firms to create circular supply agreements in which materials from end-of-life batteries re-enter new production cycles. The sophistication of the recycling process continues to advance: hydrometallurgical techniques can recover high-purity nickel, cobalt, and lithium with lower emissions, while innovative thermal and mechanical processes improve efficiency and reduce waste.

Yet recycling alone cannot support a closed-loop battery ecosystem in the near term. Most EV batteries will not reach end of life until the 2030s, meaning that recycling cannot offset surging demand in the 2020s. In the short term, recycling will rely heavily on manufacturing scrap—offcuts and defective cells from gigafactories—which provide a steady stream of material. This dynamic underscores one of the fundamental challenges of circularity: the time lag between when products are made and when they return as recyclable inputs. Effective closed-loop systems must account for these temporal gaps, ensuring stable supply during the transition period.

Designing products for disassembly is another pillar of circularity. Many modern devices—from smartphones to EV motors—are engineered for compactness, not recyclability. Adhesives, welded components, miniaturized designs, and tightly integrated architectures make it extremely difficult to recover materials without destroying parts or incurring high costs. Shifting toward design-for-repair and design-for-recycling principles will require coordinated efforts across the entire value chain. Companies must rethink product architectures. Regulators must provide incentives or mandates. Consumers must adapt to new models of ownership, repair, and reuse.

Beyond batteries, circular economy strategies extend to magnets, electronics, solar panels, wind turbine components, and industrial catalysts. Rare earth magnets, for example, can be recovered from end-of-life motors and generators. Although technically challenging, magnet recycling could reduce dependence on rare earth mining and processing, which are heavily concentrated in specific countries. Electronics recycling, meanwhile, offers opportunities to extract copper, gold, silver, palladium, indium, and gallium from complex waste streams. As digital devices become more ubiquitous, the volume of electronic waste grows, making it both an environmental hazard and a potential source of high-value materials.

Solar panel recycling is emerging as another critical area. Photovoltaic systems contain silver, copper, silicon, and polymers that can be reused if properly separated. Wind turbines, especially their nacelles and magnets, offer similar opportunities. Circularity in renewables is essential to ensuring that clean-energy technologies do not create new waste crises decades from now. Countries investing heavily in renewable infrastructure must plan for its end-of-life management before the first generation of equipment begins to retire.

Closed-loop systems also rely on new business models. Companies are exploring leasing frameworks for batteries, motors, or entire systems, allowing them to retain ownership of materials and ensure recovery at end of life. Moblility-as-a-service models could facilitate circularity by centralizing fleet management, maintenance, and recycling. Extended producer responsibility (EPR) laws—already common in Europe—require manufacturers to take back products at end of life, further encouraging circular designs and internal recycling programs.

Governments play a fundamental role in enabling circular systems. Regulatory frameworks such as the EU Battery Regulation mandate recycled content targets, traceability, and sustainability standards. In the United States, tax incentives and grants support recycling infrastructure. In Asia, several countries are developing industrial symbiosis zones where waste from one process becomes feedstock for another. Meanwhile, trade policies increasingly link market access to responsible sourcing and recycling commitments.

Internationally, circularity intersects with global justice. Developing countries often bear the environmental burden of informal recycling, where hazardous processes are carried out without proper safeguards. Circular systems must therefore ensure that recycling is not outsourced to vulnerable communities. Formalizing recycling sectors, improving labor standards, and ensuring safe processes are essential to preventing exploitation. Moreover, circular strategies must consider the economic implications for resource-dependent countries. If recycling reduces global demand for mined materials, nations reliant on mineral exports could face significant revenue losses. Policymakers must balance circularity goals with equitable development strategies.

Despite its promise, the circular economy faces multiple barriers. Technological limitations remain, particularly for complex materials or low-grade waste streams. Economic obstacles include high capital costs for recycling plants, fluctuating prices for metals, and competition with cheaper primary materials. Policy inconsistencies create uncertainty, especially in jurisdictions without clear recycling mandates. Consumer behavior—such as rapid device turnover or limited participation in collection programs—further complicates efforts.

Yet the long-term trajectory is unmistakable. Circular economy principles will become increasingly central to global mineral strategies. As extraction becomes more challenging and more contested, recycling and circularity offer a pathway to resilience. Closed-loop systems reduce the geopolitical risks associated with concentrated supply chains, stabilize material flows, and align industrial activities with sustainability goals. They also offer opportunities for innovation, new industries, and job creation in fields such as advanced materials, chemical processing, and digital traceability.

Circularity is not simply a technical challenge; it is a systemic transformation. It requires reimagining how products are designed, how supply chains operate, how regulations are implemented, and how societies value resources. It challenges assumptions embedded in decades of industrial practice. But it also reflects a deeper philosophical shift: seeing materials not as disposable commodities but as assets to be stewarded through continuous cycles of use.

Long-term trajectories for the mineral system are now being shaped by choices made under uncertainty. Different paths point toward stability, fragmentation, or disorder, depending on how technology, policy, and coordination evolve. The future of material security will be defined less by forecasts than by the strategic decisions taken today.

CHAPTER 46 – 2040 SCENARIOS: OPTIMISTIC, PESSIMISTIC, AND CHAOTIC FUTURES

Imagining the world of 2040 requires confronting uncertainty, complexity, and geopolitical tension. The trajectory of critical minerals will shape the energy transition, industrial competitiveness, global security, and environmental stability. While no scenario can predict the future with precision, three broad pathways—optimistic, pessimistic, and chaotic—capture the range of possible outcomes. Each scenario reflects the interactions of markets, technology, geopolitics, governance, and natural ecosystems. Each illustrates the stakes of choices made today. And each reveals how the mineral foundations of the modern world could strengthen or destabilize the global order.

The optimistic scenario envisions a world that manages to expand mineral supply sustainably, deploy technological innovation effectively, and build resilient global governance systems. It is not a utopia—mining continues, geopolitical tensions persist, and competition remains fierce. But states, companies, and societies manage to align their interests sufficiently to avoid severe shortages or widespread instability.

In this future, massive investment in recycling and circular economy systems reduces dependence on primary mining for several key materials. By 2040, recycling of lithium-ion batteries becomes mature and efficient, supplying a significant share of global demand for lithium, nickel, and cobalt. Rare earth magnet recycling expands rapidly, supported by robust take-back programs and specialized metallurgical facilities. Digital product passports guide consumer behavior and help manufacturers recover valuable materials from electronics, vehicles, and industrial equipment. The gap between extraction and end-of-life recovery narrows, reducing pressure on ecosystems and cutting emissions from mining and refining.

Technological innovation also plays a transformative role. Sodium-ion batteries achieve commercial competitiveness in grid storage and low-range electric vehicles, easing pressure on lithium and nickel supply chains. Silicon-rich and lithium-metal anodes improve performance while reducing reliance on graphite. Rare-earth-free magnet technologies mature, becoming viable for a wide range of applications. Solid-state batteries achieve large-scale deployment, reducing cobalt usage. Advanced semiconductor materials diversify beyond gallium and germanium, making supply chains more resilient.

Deep-sea mining remains largely restricted due to environmental concerns and cautious governance mechanisms. Biomining and geothermal brine extraction scale significantly, providing low-carbon sources of key minerals. Urban mining becomes a cornerstone of industrial strategy.

Geopolitically, the world avoids bifurcation into rigid blocs. Though competition between major powers continues, strategic cooperation emerges through shared interests in climate stability and supply security. International frameworks evolve to regulate mineral supply chains, enforce environmental standards, and mediate disputes. The International Seabed Authority strengthens its scientific review processes. Regional alliances support infrastructure investments in resource-rich nations, offering alternatives to exploitative or extractive models. Transparency initiatives reduce corruption and improve revenue management in mineral-producing countries.

This optimistic scenario requires unprecedented coordination, major technological advances, and sustained political will—but it is grounded in plausible trends already taking shape. The energy transition proceeds with fewer disruptions than expected. Industrial ecosystems grow more resilient. The mineral world becomes more sustainable, though still complex and competitive.

The pessimistic scenario, by contrast, imagines a world where mineral supply fails to keep pace with demand, technological substitution progresses slowly, and governance systems remain fragmented. In this future, structural shortages emerge, driving up costs, slowing the energy transition, and intensifying global inequities.

By the late 2020s and early 2030s, the clean-energy transition accelerates faster than supply chains can adapt. EV sales surge beyond expectations, grid-scale storage proliferates, and solar and wind deployments multiply. But new mines take a decade or more to come online. Permitting delays, protests, environmental concerns, and limited financing slow capacity expansion. Several major projects are stalled due to political instability, community resistance, or regulatory disputes.

As a result, chronic shortages of lithium, nickel, graphite, and rare earths appear. Prices skyrocket, then collapse, then surge again, creating instability for manufacturers and consumers. Some automakers delay electrification targets. Renewable energy projects face rising equipment costs. Nations scramble to secure long-term supplies, often entering exclusive agreements that fragment global markets.

Technological breakthroughs fail to materialize at scale. Sodium-ion batteries remain limited to niche markets. Solid-state batteries face persistent manufacturing challenges. Rare-earth-free magnets remain technically inferior. Semiconductor materials continue to depend on gallium, germanium, and other vulnerable inputs.

Recycling grows, but not fast enough to offset primary supply shortfalls. Collection systems remain uneven. Many regions lack formal recycling infrastructure. Informal sectors in developing countries continue to handle e-waste under hazardous conditions, causing environmental damage and health risks.

Geopolitically, tensions deepen. China strengthens its dominance in refining and midstream processing. Western nations attempt decoupling but struggle to build parallel supply chains. Export controls and countersanctions become more frequent. Resource nationalism intensifies as mineral-rich states impose strict requirements, renegotiate contracts, or pursue domestic value-add strategies that slow global supply. The global South gains leverage, but governance challenges lead to inefficiencies and underinvestment.

Environmental and social impacts worsen. Expedited mining projects cut corners. Fragile ecosystems suffer irreversible damage. Local communities increasingly resist mining expansion, leading to conflict. The energy transition slows substantially, preventing many countries from meeting climate goals.

Under this pessimistic trajectory, the mineral foundations of the clean-energy transition become a bottleneck that limits technological progress and exacerbates geopolitical risk.

The chaotic scenario is neither a smooth decline nor a balanced equilibrium—it is a world of unpredictable shocks, cascading crises, and nonlinear changes. In this future, mineral markets become destabilized by a convergence of climate disasters, political conflict, technological disruptions, and abrupt policy swings.

Climate impacts intensify dramatically by the 2030s. Extreme weather events disrupt mining operations worldwide: floods halt copper production in South America, droughts cripple lithium brine extraction, cyclones damage nickel and cobalt facilities in Southeast Asia, and heatwaves strain refineries and smelters. Logistics networks are repeatedly disrupted as storms damage ports, railways, and shipping routes.

Conflicts flare in key resource regions. Political instability in the Democratic Republic of Congo disrupts cobalt supply for prolonged periods. Tensions between China and the United States escalate, prompting aggressive export controls, retaliatory sanctions, and widespread supply chain fragmentation. Territorial disputes over deep-sea mining zones escalate into diplomatic or even military confrontations. Resource nationalism in several countries leads to sudden export bans or nationalization of assets.

Technological progress becomes erratic. Some breakthroughs occur—such as unexpected advances in storage or magnet technologies—but they emerge unevenly and unpredictably. Meanwhile, cyberattacks on mineral infrastructure, refineries, or trading platforms create further instability in global markets. Financial speculation intensifies as investors hedge against chaos, triggering extreme price swings.

Governance systems fracture. International institutions fail to adapt. Countries abandon cooperative frameworks in favor of unilateral strategies. Trust erodes. Supply chains splinter into regional clusters with little coordination. Black markets for minerals thrive. Illegal mining expands, causing environmental disasters. The clean-energy transition becomes turbulent and uneven, characterized by rapid spurts of progress followed by abrupt shortages and setbacks.

In this chaotic world, the mineral economy becomes a source of instability rather than resilience. Climate goals slip further out of reach. Inequality widens as some nations secure stable mineral access while others face scarcity. The global system becomes increasingly difficult to manage.

These scenarios are not predictions—they are lenses for understanding the choices, risks, and opportunities ahead. The world of 2040 will likely contain elements of all three: technological breakthroughs tempered by geopolitical tension, sustainability gains offset by ecological strain, cooperation coexisting with competition.

The mineral future will be shaped less by inevitability than by governance. Standards, policy architectures, and international coordination will determine whether competition hardens into fragmentation or evolves into managed interdependence. In this domain, institutional design becomes a decisive lever of stability and shared prosperity.

CHAPTER 47 – GOVERNANCE, STANDARDS, AND GLOBAL POLICY ARCHITECTURES

The future of critical minerals will not be determined solely by geology, technology, or markets. It will be shaped—perhaps decisively—by governance: the rules, institutions, standards, treaties, and policy systems that structure the extraction, processing, trade, and reuse of essential materials. In a world where critical minerals underpin everything from electric vehicles and renewable energy to semiconductors and defense technologies, robust governance is not a peripheral concern. It is the backbone of stability, sustainability, and international security. Without coherent governance frameworks, the global scramble for resources risks becoming fragmented, exploitative, and destabilizing. With the right architectures in place, however, the transition to a mineral-intensive world can proceed with transparency, fairness, and resilience.

Governance of critical minerals begins at the national level. Countries define their own extraction laws, environmental standards, labor protections, revenue-sharing models, and permitting systems. Yet huge disparities exist among jurisdictions. Nations such as Canada, Australia, and Norway have strong regulatory systems that balance development, environmental protection, and community rights. Others struggle with weak institutions, corruption, or political instability. These gaps create uneven playing fields and influence where companies choose to invest. They also affect how benefits are distributed within countries—whether mineral wealth funds education, health, and infrastructure, or flows into the hands of elites.

For resource-rich nations seeking to harness mineral wealth responsibly, governance must address several core challenges: enforcing environmental standards, ensuring meaningful community consultation, creating transparent fiscal regimes, and strengthening oversight institutions. Local participation is especially critical. Indigenous communities and rural populations often bear the burdens of extraction while receiving few of the benefits. Governance frameworks that respect land rights, cultural values, and social consent—through mechanisms such as free, prior, and informed consent (FPIC)—are essential to preventing conflict and building long-term trust.

National governance alone, however, cannot manage the global nature of critical minerals. Supply chains span continents. Processing is concentrated in a handful of countries. Markets are shaped by geopolitical rivalry and economic interdependence. This makes international cooperation indispensable. Yet current global governance structures remain fragmented. No equivalent to the International Energy Agency or OPEC exists for critical minerals. Instead, disparate initiatives—some governmental, others private, others multilateral—attempt to fill the gaps.

One of the most important multilateral frameworks is the Extractive Industries Transparency Initiative (EITI), which requires participating countries and companies to disclose payments, contracts, and beneficial ownership information. While EITI has improved transparency in many jurisdictions, it remains voluntary and does not cover all minerals or countries. Its impact depends on political will. Still, it provides a foundation for combating corruption and ensuring that mineral revenues benefit citizens.

Newer alliances reflect geopolitical realignments. The Minerals Security Partnership (MSP), led by the United States with partners such as the EU, Japan, Australia, Canada, and South Korea, aims to secure resilient supply chains through coordinated investment and shared standards. The MSP is not a treaty-based institution but an alignment of interests among industrialized nations seeking to diversify away from dominant suppliers and reduce systemic vulnerabilities. Its projects include joint financing of mining and processing facilities, harmonized ESG requirements, and shared data on supply chain risks.

The European Union has introduced its own governance architecture through the Critical Raw Materials Act, which sets targets for domestic extraction, processing, recycling, and diversification. It includes mandatory due diligence, permitting reforms, and monitoring mechanisms. The Act positions the EU as a global leader in regulatory governance, influencing standards beyond its borders by shaping the requirements for access to its markets.

At the global level, the International Seabed Authority (ISA) regulates mineral activities in international waters. Its mandate is unique: to balance resource development with environmental protection in one of the world's least understood ecosystems. Yet the ISA faces criticism for insufficient scientific rigor, limited transparency, and governance structures dominated by political bargaining. As countries debate whether to proceed with deep-sea mining, the ISA's ability to enforce strong environmental safeguards will shape the future of ocean governance and set precedents for emerging extraction frontiers.

Environmental, social, and governance (ESG) frameworks—developed by corporations, investors, civil society groups, and standard-setting bodies—serve as another layer of governance. These frameworks establish expectations for responsible sourcing, labor conditions, environmental performance, and community engagement. Companies increasingly rely on certification schemes, independent audits, and sustainability standards to demonstrate compliance. Examples include the Initiative for Responsible Mining Assurance (IRMA), the Responsible Cobalt Initiative (RCI), and the Aluminium Stewardship Initiative (ASI). While these standards improve accountability, they remain voluntary and vary widely in their requirements and enforcement.

Digital tools are strengthening governance in important ways. Traceability systems—using blockchain, digital passports, or advanced tagging technologies—allow companies to track materials from mine to factory to consumer. These systems help prevent fraud, ensure compliance with regulatory requirements, and provide consumers with transparent information about the origins of products. The European Union's plans for digital product passports will set a global precedent, requiring detailed disclosure of material composition, carbon footprint, and recycling pathways. If adopted globally, digital traceability could transform how critical mineral supply chains are monitored and regulated.

International financial institutions also shape governance through conditional financing. The World Bank, the African Development Bank, and the Asian Development Bank fund infrastructure and capacity-building projects that influence mining governance, environmental protections, and local development. Their lending criteria often include safeguards related to human rights, environmental impact, and transparency. Yet these institutions cannot meet global demand for financing the mineral transition. As a result, emerging lenders—especially from China and the Gulf—

play an increasingly influential role, often with different conditions and priorities.

Geopolitics complicates governance. Rivalries between major powers influence investment decisions, regulatory approaches, and bilateral agreements. Resource nationalism, supply chain weaponization, and strategic stockpiling create tensions that undermine collaborative governance. The fragmentation of global markets into competing blocs—with parallel supply chains, different standards, and competing alliances—poses a major challenge to building a cohesive global governance framework.

Still, opportunities exist for cooperation. Climate goals create shared interests among nations that might otherwise compete. No country can achieve energy transition targets without reliable access to critical minerals. Shared vulnerability can drive shared solutions. Multilateral forums—such as the G20, the OECD, and the UN Environment Programme—can expand their mandates to support dialogue on critical minerals, harmonize standards, and coordinate investment. The International Energy Agency's analytical role could expand to include more systematic monitoring of mineral supply chains.

By 2040, several governance pathways are possible. In a best-case trajectory, countries adopt coordinated standards, enforce robust environmental protections, and invest collectively in resilient and ethical supply chains. Transparency improves, corruption declines, and resource-rich nations benefit more equitably from mineral wealth. In a worst-case scenario, governance remains weak, fragmented, or politicized. Competition escalates, environmental harm increases, and communities bear the brunt of extraction without receiving fair compensation.

The stakes are enormous. Governance will determine whether the mineral-intensive future is one of resilience and sustainability—or one of conflict, inequality, and ecological damage. The clean-energy transition depends not only on technological innovation or market forces but on the institutions that guide behavior, mediate interests, and uphold accountability.

An era is emerging in which minerals shape power, but governance determines its direction. Decisions taken across policy, industry, science, and diplomacy are already setting the contours of future stability and inequality. What follows from here will reflect not scarcity alone, but the collective choices made in response to it.

CONCLUSION
CHAPTER 48 – THE INVISIBLE THREAD HOLDING THE TECHNOLOGICAL WORLD

The modern world rests on an invisible foundation. Our devices, vehicles, energy grids, communication systems, medical equipment, and military technologies all rely on a hidden lattice of materials that most people never see, never touch, and rarely think about. These materials are measured in grams, kilograms, or parts per million—yet without them, the entire technological architecture of the twenty-first century would collapse. They are the quiet enablers of progress, the silent infrastructure behind innovation, the physical backbone of the digital age. They form the invisible thread holding the technological world together.

This invisibility is not accidental. Most critical minerals lie far from the places where they are consumed. They are mined in remote deserts, tropical jungles, high-altitude plateaus, deep underground tunnels, or artisanal pits that exist far beyond the view of the average consumer. They are refined in specialized facilities, processed in industrial clusters, and embedded deep inside devices that reach our hands with no trace of the long, complex journey that brought them there. The sleek glass of a smartphone conceals dozens of minerals sourced from multiple continents. The smooth hum of an electric motor hides the rare earth magnets spinning at its core. The quiet reliability of a wind turbine is sustained by supply chains stretching across oceans and political borders.

Modern technology is celebrated for its immaterial aura—data clouds, virtual networks, artificial intelligence, software layers. Yet every digital action depends on physical materials. Without gallium, germanium, and silicon, the semiconductor revolution would not exist. Without cobalt, nickel, lithium, and graphite, the electric vehicle revolution would stall. Without copper, silver, and tellurium, renewable energy systems would grind to a halt. Even the most advanced technologies, from quantum computing to aerospace engineering, rely on metals whose names rarely appear in public discourse.

This tension between visibility and invisibility shapes the way societies understand technology. We tend to celebrate the breakthroughs but ignore the materials that make them possible. The critical minerals story reveals how deeply interconnected the world has become—and how vulnerable.

Understanding this invisible thread means recognizing that global power is no longer defined solely by military strength, economic dominance, or technological advancement. It is defined by the capacity to access, process, recycle, and innovate around strategic materials. Nations that can secure these materials build the foundations for prosperity. Nations that cannot must navigate a world of dependence and uncertainty. This reality is reshaping alliances, policies, industrial strategies, and geopolitical rivalries.

The invisible thread is also a story of fragility. Supply chains built over decades can fracture under the pressure of a pandemic, a natural disaster, a political conflict, or a single regulatory decision. A shutdown in a far-away refinery can disrupt entire industries. A diplomatic dispute can trigger export controls that ripple across global markets. When materials critical to semiconductors, batteries, or renewable technologies become scarce, the effects are felt everywhere—from automotive plants to hospitals, from power grids to defense systems.

Yet within this fragility lies opportunity. The recognition that materials matter has awakened a new era of strategic thinking. Governments are reexamining their industrial foundations. Companies are reimagining their supply chains. Investors are reassessing long-term risks. Scientists and engineers are exploring new frontiers of innovation—developing substitutes, improving recycling methods, and designing more efficient technologies. Societies are beginning to understand that the invisible thread is not just a source of vulnerability but a space for creativity and reinvention.

The invisible thread is also a human story. Behind every tonne of mined material stand workers, communities, and ecosystems. Mining can bring economic opportunity but also environmental damage, social conflict, and political instability. The global quest for strategic materials compels a renewed focus on justice, equity, and ethical responsibility. The clean-energy transition must not replicate the extractive injustices of the past. The shift to renewable systems requires not only technological innovation but moral clarity—ensuring that the benefits of progress do not come at the expense of the world's most vulnerable populations.

At the same time, the invisible thread symbolizes resilience. Humanity has always adapted to material constraints. From the Bronze Age to the Silicon Age, societies have thrived by learning to work with the resources at hand. The current period is no different, though the stakes are far higher. Climate change, digital dependence, and geopolitical complexity demand a far more sophisticated understanding of materials than ever before. The invisible thread will hold only if societies cultivate resilience—through diversified supply chains, sustainable extraction practices, circular systems, and technological innovation.

This chapter also marks a shift from analysis to synthesis. The journey through the value chains, geopolitical dynamics, technological transformations, and sustainability challenges leads to a deeper realization: the mineral world is not separate from the technological world—it is its very foundation. The global economy is not abstract—it is material. Energy security is mineral security. Technological innovation is mineral innovation. National power is mineral power.

Yet despite its importance, the mineral system is still poorly understood by the public and often underestimated by policymakers. Part of the challenge is psychological: minerals are out of sight, buried underground or hidden inside devices. Another challenge is informational: supply chains are complex, opaque, and constantly evolving. But the greatest challenge is strategic: societies must think long-term about materials at a moment when political and economic systems reward short-term decisions.

Looking ahead, the invisible thread will become increasingly visible. Scarcity, competition, and environmental challenges will force more transparency. Recycling systems will bring materials back into the public eye. Policies will mandate accountability. Digital traceability will expose the origins of every component. In time, people may come to view materials not as distant commodities but as essential elements of daily life.

The story of the invisible thread is ultimately a story of interdependence. No nation can produce everything it needs. No company can control every step of its supply chain.

CONCLUSION

No technology exists in isolation. And no transition—whether toward clean energy, digital infrastructure, or advanced manufacturing—can succeed without recognizing the material realities underpinning progress.

This chapter invites a broader reflection: the technological world may appear intangible, but it is rooted in the Earth itself. The minerals that power our societies remind us that innovation is grounded in the physical universe. They remind us that progress has limits, and that stewardship is essential. They remind us that the future must be built not only with creativity and intelligence but with humility and responsibility.

Navigating the decades ahead will require confronting scarcity without defaulting to instability. Security, resilience, and sustainability must be embedded directly into the material foundations of global power. The challenge is no longer whether minerals matter, but how societies choose to organize around their limits.

CHAPTER 49 – NAVIGATING SCARCITY, SECURITY, AND SUSTAINABILITY

The tension between scarcity, security, and sustainability defines the strategic landscape of the twenty-first-century mineral world. These three forces—each powerful on its own—interact in ways that shape national policy, corporate strategy, technological development, and global power. To navigate the decades ahead, societies must understand that scarcity is not merely a physical condition but a structural reality; security is not simply military or economic but deeply material; and sustainability is not an optional ideal but the prerequisite for long-term stability. Together, they form a triangle of constraints and opportunities that determine how nations, industries, and communities adapt to a resource-intensive technological age.

Scarcity is a growing feature of the global minerals system. It takes many forms: geological scarcity, economic scarcity, geopolitical scarcity, technological scarcity, and even environmental scarcity. Geological scarcity refers to the uneven distribution of mineral deposits. Many of the materials essential for clean energy and advanced technologies—such as cobalt, rare earths, graphite, gallium, germanium, and high-grade nickel—are concentrated in a handful of countries. These deposits often lie in places with fragile ecosystems or weak governance, increasing the risk of conflict and instability.

Economic scarcity results from underinvestment in mining and processing. Even when minerals are abundant, they may remain inaccessible because financing is insufficient, permitting is slow, infrastructure is lacking, or market conditions are uncertain. The long cycles of boom and bust discourage consistent investment, creating periodic shortages. Meanwhile, demand for minerals grows steadily, driven by electric vehicles, solar power, wind turbines, semiconductors, robotics, and digital infrastructure. The gap between demand expectations and investment realities increases the probability of structural shortages.

Geopolitical scarcity arises from political decisions—export controls, sanctions, nationalization, or strategic stockpiling—that restrict global flows. When a country capable of supplying essential minerals limits exports, scarcity becomes a tool of influence. Technological scarcity emerges when industries depend on specific materials for which no alternatives exist. Even small disruptions in these markets can derail industrial production. And environmental scarcity arises when ecosystems cannot sustain extraction without irreversible damage, forcing societies to confront the limits of resource exploitation.

Security, meanwhile, is increasingly synonymous with material security. Nations that once defined security through energy independence or military capacity must now consider the stability of mineral supply chains. The electrification of transportation, the digitalization of economic life, and the modernization of military systems all rely on materials with complex, vulnerable supply chains. A shortage of lithium can slow national electrification plans. A disruption in rare earth magnet supply can impair weapons systems and renewable energy deployment. A bottleneck in semiconductor materials can paralyze entire industries.

Material security therefore extends far beyond traditional resource policy. It requires diversification of supply, investment in domestic or allied refining, strategic stockpiles, and long-term contracts with producers. It may also require reshoring key parts of supply chains, developing alternative technologies, and building international partnerships that align industrial interests with geopolitical stability. Security in a mineral-intensive world is multi-layered—economic, technological, environmental, and geopolitical.

At the same time, sustainability cannot be an afterthought. The energy transition aims to solve the greatest environmental challenge in human history—climate change—but the materials that enable this transition carry their own environmental and social consequences. Mining can cause water depletion, pollution, biodiversity loss, and community displacement. Refining can generate hazardous waste and carbon emissions. Without strong environmental stewardship, the pursuit of clean energy risks creating new ecological crises.

Sustainability therefore redefines how scarcity and security are understood. If mining destroys water sources, it deepens scarcity. If supply chains rely on exploited labor or polluted ecosystems, they undermine social stability and long-term security. If mineral extraction accelerates carbon emissions or weakens environmental resilience, it contradicts the very purpose of the energy transition. Sustainability is not a constraint—it is a strategic advantage. Nations and companies that adopt cleaner technologies, stronger protections, and circular systems will be more resilient, more competitive, and more trusted.

Navigating the interplay of scarcity, security, and sustainability requires new approaches to policy and governance. First, governments must adopt long-term strategies that integrate industrial policy, environmental protection, and geopolitical planning. Traditional, siloed approaches are insufficient. Mineral policy must connect with climate policy, trade policy, labor policy, national security, and technology development. Countries that treat critical minerals as a short-term commodity challenge will fall behind. Those that understand minerals as a strategic pillar of the future economy will prosper.

Second, markets must adapt to the realities of mineral-intensive technologies. Price volatility, investment cycles, and supply bottlenecks cannot be resolved by market forces alone. Governments, companies, and investors must collaborate to build more resilient value chains. This includes funding for exploration, processing, recycling, and new extraction technologies; reforms to permitting; and investment in infrastructure. These actions reduce economic scarcity and strengthen national security.

Third, international cooperation is essential. No nation can achieve mineral security in isolation. Shared standards, joint investments, diplomatic partnerships, and multilateral forums are needed to manage competition and foster responsible development. Collaborative frameworks can prevent resource nationalism from spiraling into conflict and ensure that the energy transition does not exacerbate global inequality.

Fourth, sustainability must be embedded into every step of the value chain. Strong environmental regulations, ethical sourcing requirements, community engagement, and technological innovation can minimize ecological and social harm. A responsible mineral economy requires transparency, accountability, and fairness. ESG frameworks alone are not enough—they must be enforced and integrated into broader governance systems. Sustainability is not just an environmental imperative but a foundation for long-term supply stability.

Fifth, technology must play a transformative role. Reducing dependence on scarce minerals through substitution, efficiency improvements, and design innovation can alleviate pressure on supply chains. Circular economy systems can recover materials at scale, reduce waste, and provide a buffer against geopolitical disruptions. Breakthroughs in battery chemistry, magnet design, semiconductor materials, and recycling processes will shape the 2040 landscape far more than any single new mine. Innovation is the ultimate form of energy and mineral security.

Navigating scarcity, security, and sustainability also requires confronting difficult trade-offs. For example:

Should a country accelerate mining expansion to secure materials quickly, even if environmental impacts are not fully resolved?

Should governments rely on allies for critical minerals or pursue full domestic autonomy, even at higher cost?

Should societies accept higher prices for technology in exchange for more sustainable supply chains?

Should communities prioritize short-term economic benefits from mining over long-term environmental stability?

These trade-offs do not have simple answers. But the future will be determined by how leaders, industries, and societies choose to balance them.

The chapter ultimately highlights a profound truth: the mineral era is not a temporary phase—it is the structural foundation of the technological future. How nations navigate scarcity, secure materials, and pursue sustainability will define their competitive position in the decades ahead. Scarcity cannot be eliminated, but it can be managed. Security cannot be guaranteed, but it can be strengthened. Sustainability cannot be postponed, because the costs of ignoring it compound rapidly.

The final chapter will build on these insights to explore how strategic materials will shape global power—highlighting the tectonic shifts, emerging trends, and long-term dynamics that will define the world order in the age of minerals.

CHAPTER 50 – THE FUTURE OF STRATEGIC MATERIALS AND GLOBAL POWER

The story of strategic materials is, ultimately, a story about the future of global power. The minerals that flow through batteries, semiconductors, renewable energy systems, and advanced manufacturing are not just inputs in technological progress—they are levers of influence, foundations of national strength, and determinants of geopolitical stability. As the world accelerates toward electrification, digitalization, and automation, control over these materials will shape the balance of power more profoundly than oil did in the twentieth century. The future will be written not only in innovation laboratories and international summits but also in the mines, processing facilities, recycling hubs, and research centers where the material architecture of tomorrow is forged.

In the decades ahead, three forces will define the strategic landscape: technological acceleration, geopolitical polarization, and ecological constraint. Together, they create both extraordinary opportunity and heightened risk. Nations capable of aligning these forces with foresight, governance, and innovation will rise. Those that fail to adapt may find themselves strategically exposed, economically constrained, or technologically dependent.

The first force is technological acceleration. The world is entering an era where the pace of innovation in energy systems, electronics, and advanced materials is unprecedented. New battery chemistries will reshape mineral demand patterns. Rare-earth-free magnets could diminish dependency on a handful of countries. Solid-state batteries, silicon-rich anodes, sodium-ion technologies, and alternative semiconductor materials will transform industrial ecosystems. Recycling technologies will expand, turning waste into strategic assets and reducing reliance on primary extraction. Even the concept of scarcity itself will evolve as scientific breakthroughs alter what materials are considered essential.

Yet technological acceleration does not guarantee stability. It creates new dependencies even as it resolves old ones. A breakthrough that reduces demand for one mineral may increase demand for another. Substitution may lower geopolitical risk in one area while introducing vulnerabilities elsewhere. Innovation, therefore, is both a liberating and destabilizing force. The nations that thrive will be those that anticipate shifts rather than react to them, build adaptive industrial policies, and invest in research early enough to shape technologies rather than chase them.

The second force shaping the future is geopolitical polarization. The contest between major powers—especially the United States and China—will increasingly revolve around control of materials, technologies, and supply chains. This rivalry will influence alliances, trade patterns, industrial strategies, and financial flows. It will shape where manufacturing occurs, which countries gain access to new technologies, and how global standards are written.

China's dominance in processing, magnet production, battery materials, and midstream refining gives it enormous strategic leverage. Even if other nations invest in mining and recycling, they must contend with China's integrated value chains, vast industrial capacity, cost advantages, and political coordination. The United States and its partners, meanwhile, seek to rebuild industrial capabilities, diversify supply sources, and establish mineral alliances that reduce strategic dependence. Europe faces the dual challenge of securing supply while adhering to stringent environmental standards. India, South Korea, Japan, and Southeast Asian nations will play increasingly important roles as competitive industrial hubs and geopolitical intermediaries.

Resource-rich regions—Africa, South America, Central Asia—will gain strategic importance. Countries with significant mineral reserves will not remain passive suppliers; they will negotiate from stronger positions, pursue value-added industrialization, and shape alliances based on long-term national interests. The global South will become a central arena for mineral diplomacy, investment competition, and infrastructure development. The future balance of power will be influenced as much by Kinshasa, Ulaanbaatar, and Santiago as by Washington or Beijing.

The third force is ecological constraint. Climate change, biodiversity loss, water scarcity, and environmental degradation will place strict limits on how minerals are extracted and processed. Societies will increasingly refuse to accept environmental damage as the cost of technological advancement. Governments will impose stronger regulations. Communities will assert land rights and challenge extractive models. Investors will demand transparency and sustainability, not only as ethical imperatives but as risk management strategies. The ecological limits of the planet will force a rethinking of how materials are sourced, used, and circulated.

These constraints will accelerate the rise of circular economy systems, urban mining, and low-impact extraction technologies. They will push innovation toward materials efficiency, responsible mining, and closed-loop systems. Ecological limits will not halt the energy transition—but they will shape its speed, cost, and political legitimacy.

In this complex landscape, several long-term dynamics stand out. First, material supply chains will become strategic assets akin to digital infrastructure or military alliances. Nations will build stockpiles, diversify processing capabilities, and invest in strategic partnerships. Industrial policy will gain importance. Governments will play a far more active role in shaping mineral ecosystems, from funding research to supporting extraction projects to guiding corporate behavior through standards and incentives.

Second, global power will become increasingly multipolar. Rather than a single hegemon dominating mineral supply, a constellation of actors will shape different segments of the value chain. China may retain dominance in processing, but the United States could lead in recycling technologies, Europe in sustainability standards, Japan in advanced materials, and India in refining expansions. Africa and Latin America may assert control over upstream resources, while Southeast Asia emerges as a hub for manufacturing and midstream processing. Power will be distributed across networks rather than concentrated in single centers.

Third, the line between economic and security policy will blur. Minerals will become instruments of leverage, negotiation, and deterrence. Export controls, sanctions, investment screening, and industrial subsidies will become common tools of statecraft. Nations will use material access to gain strategic advantage, reward allies, or punish adversaries. Companies will face unprecedented scrutiny over sourcing, partnerships, and supply chain vulnerabilities.

Fourth, innovation will become the most important determinant of long-term power. Countries that lead in materials science, chemical engineering, magnet technology, battery innovation, and recycling will shape the global mineral order. Scientific leadership will define technological independence, industrial competitiveness, and geopolitical strength. The race for innovation will be as important as the race for resources themselves.

Finally, ethical considerations will rise in importance. Societies will demand that the materials enabling the clean-energy transition be sourced without injustice, exploitation, or environmental destruction. Supply chains that rely on harmful practices will lose legitimacy and face regulatory, consumer, and investor pushback. Ethical leadership in minerals will become a form of soft power, shaping alliances and influencing global norms.

The future of strategic materials and global power is therefore neither predetermined nor uniform. It will be shaped by choices—choices about governance, investment, innovation, diplomacy, and sustainability. These choices will determine whether the mineral era becomes a source of cooperation or conflict, stability or instability, collective progress or unequal struggle.

This book has traced the mineral foundations of the modern world—from geology and supply chains to geopolitics and future scenarios. The conclusion is clear: critical minerals are not merely resources; they are the architecture of the future. The nations that recognize this, act upon it with foresight, and balance scarcity, security, and sustainability will shape the coming era. Those that fail to adapt risk losing technological sovereignty, economic competitiveness, and geopolitical influence.

The mineral age is here. Its challenges are immense, but so are its possibilities. The path forward requires intelligence, responsibility, innovation, and cooperation. The future of global power will belong to those who understand that the next great strategic frontier is not only digital or military—it is material.

ABOUT THE AUTHOR

The Top-Tier Education Program is a publishing initiative dedicated to designing and producing advanced strategic books for entrepreneurs, managers, professionals, and individuals seeking tools to navigate contemporary transformation. The Program focuses on developing content that combines theoretical depth and practical applicability, leveraging a transdisciplinary and systemic approach. Each volume is designed as a structured learning platform, offering conceptual maps and toolkits to orient readers in challenges such as leadership, strategy, innovation, sustainability, and organizational culture. The mission is to deliver high-quality, accessible resources that empower readers to understand the hidden structures shaping the emerging world order, without making claims regarding the individual authorship of the content.

www.ingramcontent.com/pod-product-compliance
Lightning Source LLC
LaVergne TN
LVHW022048200926
848481LV00003B/836